T0181128

Additive Manufacturing Change Management

Continuous Improvement Series

Elizabeth A. Cudney and Tina Kanti Agustiady

For more information about this series, please visit: www.crcpress.com/
Continuous-Improvement-Series/book-series/CONIMPSER

Additive Manufacturing Change Management
Best Practices

David M. Dietrich, Michael Kenworthy,
and Elizabeth A. Cudney

CRC Press
Taylor & Francis Group
Boca Raton London New York

CRC Press is an imprint of the
Taylor & Francis Group, an **informa** business

CRC Press
Taylor & Francis Group
6000 Broken Sound Parkway NW, Suite 300
Boca Raton, FL 33487-2742

International Standard Book Number-13: 978-0-367-15207-9 (Hardback)
International Standard Book Number-13: 978-1-138-61175-7 (Paperback)

Library of Congress Cataloging-in-Publication Data

Names: Dietrich, David M., author.
Title: Additive manufacturing change management : best practices / authored by David M. Dietrich, Michael Kenworthy, and Elizabeth A. Cudney.
Description: Boca Raton : Taylor & Francis, 2019. | Series: Continuous improvement series | Includes bibliographical references.
Identifiers: LCCN 2018047990| ISBN 9781138611757 (paperback : alk. paper) | ISBN 9780367152079 (hardback : alk. paper) | ISBN 9780429465246 (e-book : alk. paper)
Subjects: LCSH: Three-dimensional printing. | Manufacturing processes—Automation.
Classification: LCC TS171.95 .D54 2019 | DDC 621.9/88—dc23
LC record available at https://lccn.loc.gov/2018047990

Visit the Taylor & Francis Web site at
http://www.taylorandfrancis.com

and the CRC Press Web site at
http://www.crcpress.com

Contents

Foreword

Additive manufacturing (AM) is an exciting new manufacturing technology, and it has captured the imagination of tens of thousands of people across the globe. It has brought young people into manufacturing like never before. A student graduating this year has likely been exposed to 3D printing in high school, middle school, or perhaps even at the primary school level. It is up to the professionals in the industry to keep these young people engaged and involved to make all products better.

We see many companies with aspirations of jumping immediately in to AM because they want to make AM parts as fast as possible. This action, however, is a mistake. We see AM as a series of technologies, versus "a single technology," which suggests that understanding the concept is likely to ensure a higher degree of success. Taking a "maturity model" approach, we would recommend starting with tools and fixtures not because it's easy, but because it can save money and it helps build a learning culture in AM. As with any technology maturation, there should be a building block approach to reduce risk and improve learning relative to the requirements.

This book is as much about the importance of corporate culture and its ability to promote innovation as it is about AM. AM is a victim of its own success; everyone wants to be seen as involved in AM, but the reasons can be convoluted. Just as "greenwashing," we see corporations using AM to fill their annual reports and recruit the right talent. Like most of manufacturing, AM is hard and requires persistent care and attention to detail. It also requires paperwork, procedures, and specifications. It's manufacturing, yes, but AM has largely evolved from prototyping, which has an inverse requirement need.

With a mighty organization come both strengths and weaknesses. As I believe any team can be high performing when confronted with the right circumstances, AM can also be successful when used in the right environment. AM is a team sport and, as the authors point out, requires different disciplines to come together to solve a harder problem. Everything in today's world is a sourcing or supply chain risk, or an engineering, and certification challenge, and word travels quickly. An organization must allocate strong attention and focus to their AM strategy development. While you may have the technical assets to do the work, as with any army, it travels on its stomach. You will also need supply chain, procurement, and all of the corporate functions in the convoy.

The authors clearly make the case that teams willing to commit to the journey can be successful. What is most compelling about this book is that rather than build on the hype for AM, they diagnose the issues that are likely to take place within any complex organization. AM has to do everything legacy methods can do, and it must do it both better and cheaper.

The best companies accept innovation as a cultural journey that prepares them to be ready for new technologies as they arrive. Constant encouragement for training of employees promotes personal and professional growth which drives engagement. It is a bright future for the next generation of manufacturing professionals. We have only begun to scratch the surface of what this technology even means, much less what it can do.

John E. Barnes
Founder and Managing Director
The Barnes Group Advisors.

Preface

Some years ago, I was on the road visiting customers and telling them about what we were doing with AM and how it had transformed our product line capabilities. I came to realize that not only had it changed the product, but also fundamentally changed how we must collaborate to effectively leverage the new technology and drive stakeholder acceptance on both sides. It is fair to characterize most of our supplier–customer interactions at the time as largely transactional engineering – we receive a request for quote with a design specification and box in which to fit an iteration of previously provided product.

This kind of relationship makes little sense once you equip teams with a toolset assembled specifically to invalidate the assumptions underpinning said box. I became convinced that attempting to "do AM" without the customer-in-the-loop is a grave mistake, and the line between my team and the customer's engineering team needed to blur. Most of the time, however, it felt like we were IBM introducing the Simon smartphone in 1992 (impressively, it had a stylus-operated touchscreen and email/fax capability). Were we also trying to sell something too advanced for its day? Maybe the customer really did want just another box because she had much bigger problems on her hands and didn't really have time for this new AM stuff.

Out of these many customer conversations, I gradually assembled a multipage document that eventually served as a strategic roadmap for assessing an organization's AM maturity and determining winning strategies to get better at it together. This was key to moving my own approach to implementing AM from a primarily tactical, project execution-oriented approach to one with its guiding principle as enabling customer success. Once I looked at individual projects in this context, it became apparent where resources should and should not be applied to advance the state of AM with both internal and external customers.

I used this document for a few years, gradually adding to it and modifying it where my knowledge of AM realities had grown. I wasn't surprised to find that Dave had many similar experiences, but they were almost too similar. As we shared and debated these key events and outcomes, and how alternative timelines could have played out, the organizational root causes for lack of acceptance became clearer to both of us. This book *Additive Manufacturing Change Management: Best Practices* is a culmination of two rather exciting and, at times, baffling journeys with AM, and it summarizes the useful tools and lessons we have learned along the way. Of course, there is no reason to invent entirely new tools, where the existing ones satisfy the need and can be cleverly adapted to the challenge. Beth has been integral to this process as a well-published author and expert in the field of Lean Systems and Six Sigma. She has brought a wealth of knowledge in these process improvement systems and suggested ways to apply them directly to AM challenges facing industry today. We all sincerely hope that body of knowledge can help you to avoid the worst of the mistakes and, in some small way, thereby accelerate the entire AM industry.

AM is an emerging technology that is poised to alter the state of manufacturing as we know it. Aerospace and healthcare are the predominant fields that have taken advantage of this technology to shorten development lead time, increase product performance, and reduce costs of new products. This emerging technology requires a specific approach to technology, supplier, and enterprise communication development. However, organizations often struggle with creating a strategy both internally and for their supply chains, particularly for new manufacturing technologies. Within large companies, it is challenging to assess emerging technologies with respect to culture and barriers/risk.

Further, executive management within industry are bombarded by clever marketing from their competitors showcasing innovative design solutions leveraged through AM. Therefore, executive management tasks their management/project management teams within companies to figure out how to utilize AM within their own company. Often clueless on how to approach this new problem, managers start learning about AM from subject matter experts and become frustrated and overwhelmed at the highly technical information regarding the technology without a clear understanding of how it should be implemented.

The goal of this book is to provide general guidelines to help executive management successfully implement AM within their organization. The book is organized into four main sections. Section I provides a brief overview of AM and potential AM uses in industry. Section II discusses implementing new technologies requires understanding the impact of disruptive technology, which must be addressed prior to adoption. Case studies presenting war stories of industry implementations are provided to illustrate potential hurdles and how they were overcome. Next, as with any new technology, there are barriers. Identifying and proactively addressing these barriers increase the potential for successful implementation. These barriers and potential solutions are discussed in Section III.

Finally, best practices with respect to change management are provided. Therefore, Section IV discusses how to create and manage the necessary culture change. Finally, methods to sustain change are provided. With sincere humility, we hope that you can apply the content of this book to your company's AM journey.

David M. Dietrich, PhD
Michael Kenworthy
Elizabeth A. Cudney, PhD

Acknowledgments

David M. Dietrich, PhD

I would like to first thank my wife, Nancy, and children, Ally, Drew, Grant, and Zach, for allowing me the time to write this book. I'd also like to thank my parents for all their help over the years. Many thanks to our publisher, Taylor & Francis Group, for providing us the chance and patience to release this work. I'd also like to give a shout out to all my incredibly talented coworkers who I served in the corporate aerospace trenches with over my career; Mike, Amy, JG, Kevin, Pete, and many, many more. I've been fortunate to work with so many smart people through the years... and JG too. Also, many thanks to my managers for providing valuable lessons for this book. Good and bad, you've taught me a great deal. Lastly, I'd also like to express my gratitude to my co-authors and the various works of others in the AM industry who have given permission for us to use their material in the book.

Michael Kenworthy

It's a difficult thing to write a book like this when you're quite sure that it was only just moments ago that the gunshot signaling the start of the marathon rang out. It's also difficult to do this kind of thing without a generous portion of support and understanding from family members – I love you Sandra, Matthew, Timothy, and Kimberly. I had a healthy respect for authors before, but like many activities, actually doing it teaches you so much more about how challenging the process can be on those involved. I'm grateful that I was able to experience this with my co-authors who have been through it before.

Of course, I must thank all the exceptionally talented people I have worked with over my engineering career with MIT, General Electric, Honeywell, and now Divergent 3D. To those who have coached and inspired me to be always learning and growing, questioning assumptions, and navigating the pathways to success as a team – you know who you are. It was rarely easy, but what would be the fun in that anyway?

Elizabeth A. Cudney, PhD

Thank you to my incredibly supportive family for always believing in me. To my husband, Brian, you always push me to follow my passion. To our children, Caroline and Josh, you inspire me to continue learning and being a better person. Thank you to my sweet Wesley, Riley, and Winnie who are the best company when I write. I have the pleasure and honor to work with such talented individuals who help me grow personally and professionally.

Authors

David M. Dietrich, PhD, has 15 years' experience in AM within aerospace. Currently, he is the Additive Manufacturing Engineering Design Fellow for Honeywell Aerospace. In this role, he develops designs for AM technologies for use in aerospace production. In addition, he worked for 12 years for the Boeing Company researching AM technologies for both polymers and metals in St. Louis. During the last 2 years of employment at Boeing, he was an ORNL – Industrial Fellow embedded at Oak Ridge National Laboratory Manufacturing Demonstration Facility. Dr. Dietrich has more than 18 years of manufacturing industry experience that includes both production and research with expertise in the areas of AM, Quality Engineering, Six Sigma, and Lean production.

Academically, he's held adjunct faculty roles with the University of Tennessee-Knoxville and Missouri University of Science and Technology. He is currently an Industrial Advisory Board member for Arizona State University – Polytechnic and regularly mentors senior engineering students at ASU.

Dr. Dietrich earned his PhD in Engineering Management from the Missouri University of Science and Technology. His dissertation research aligned with the development of AM supply chains for aerospace. He also earned a Master of Manufacturing Engineering from Missouri University of Science and Technology and MBA from Maryville University of St. Louis. In addition, he earned a Bachelor of Science from Murray State University. In 2011, he was awarded the Outstanding Young Manufacturing Engineer distinction from the Society of Manufacturing Engineers (SME). He has presented at more than 20 conferences in the field of AM, holds 11 U.S. patents and published several papers on the subject.

Michael Kenworthy is currently the Vice President of Additive Manufacturing (AM) at Divergent 3D, a start-up company focused on disrupting traditional manufacturing with additive at its core. Recently, he was the Chief Engineer for AM at Honeywell Aerospace helping the team with implementation of the technology and enabling processes across a broad product portfolio of turbine engines, auxiliary power units, and other mechanical systems such as pneumatic controls. Previously, as a Technology Program Leader while at General Electric – Aviation, he held Controlled Titles for several technology areas and focused on disrupting conventional products. Earlier in his career, he worked with various other novel manufacturing technologies, new material development, and products for aerospace and industrial applications such as ignition systems and turbine engine pressure, temperature, and clearance sensors.

Mr. Kenworthy earned a BS in Mechanical Engineering from the Massachusetts Institute of Technology. Later, he earned his Master of Engineering in Management from the University of Wisconsin – Madison. He has presented at numerous conferences on such topics as Design for Additive Manufacturing and Engineering Change Management and holds several granted U.S. patents, many of them enabled by AM.

Elizabeth A. Cudney, PhD, is an Associate Professor in the Engineering Management and Systems Engineering Department at Missouri University of Science and Technology. She earned her BS in Industrial Engineering from North Carolina State University, Master of Engineering in Mechanical Engineering and Master of Business Administration from the University of Hartford, and her doctorate in Engineering Management from the University of Missouri – Rolla.

In 2018, Dr. Cudney was elected as a Fellow of the Institute of Industrial and Systems Engineers (IISE). In 2017, she received the 2017 Yoshio Kondo Academic Research Prize from the International Academy for Quality (IAQ) for sustained performance in exceptional published works. In 2014, Dr. Cudney was elected as an American Society for Engineering Management (ASEM) Fellow. In 2013, she was elected as an American Society for Quality (ASQ) Fellow. In 2010, Dr. Cudney was inducted into the International Academy for Quality. She received the 2008 ASQ A.V. Feigenbaum Medal. This international award is given annually to one individual "who has displayed outstanding characteristics of leadership, professionalism, and potential in the field of quality and also whose work has been or, will become of distinct benefit to mankind." She also received the 2006 SME Outstanding Young Manufacturing Engineering Award. This international award is given annually to engineers "who have made exceptional contributions and accomplishments in the manufacturing industry." Dr. Cudney has published 7 books and more than 70 journal papers. She holds eight ASQ certifications, which include ASQ Certified Quality Engineer, Manager of Quality/Operational Excellence, and Certified Six Sigma Black Belt, among others.

Introduction

Today, you can hardly go a week without reading about some company announcing it is "industrializing additive manufacturing." Ready to follow fast, engineering leaders everywhere are being tasked to develop an AM strategy. Perhaps this leader is even given a significant budget right from the start without any input as to the needed amount and is reviewing the team's requests to begin buying various equipment. Industrializing AM is a complex undertaking and involves far more than acquiring machines and placing them in a building with a few newly trained operators and engineers. As the leader, you must ensure that the team is maintaining a broad view of the value AM can bring to the business.

Those looking at the deployment of AM strategy on a selective basis such as part cost or meeting a pressing obsolescence or sustainment need will surely miss the true array of advantages it offers. The team will need to rethink the entire production process from design to field service and how AM can and should change each step. You are responsible for maintaining the strategic view even as the day-to-day tactical concerns threaten to overwhelm the team. Ensure that these system-level changes are implemented, by influencing the key stakeholders in your organization, or this undertaking will be resisted for this is the nature of change.

Seek out constructive confrontation and continually question your assumptions. Be prepared to alter the plan and secure buy-in from the team. Do not accept short-term compromises that will damage the long-term health of the AM transformation you are leading. As famously advised by Andy Grove, "Business success contains the seeds of its own destruction," "Success breeds complacency. Complacency breeds failure," "Only the paranoid survive." We are confident that the following questions can help you to define how to start, refine your thoughts, and build a winning strategy unique to your business's needs.

- Looking across my organization, what current products are most in need of disruptive innovation? Are there any part families where AM can confer a competitive advantage?

- What tools will we use to form the initial strategic vision and at what frequency should it be updated? Who are the stakeholders that should participate in this process?
- What resources are available to me to ensure a robust intellectual property (IP) plan that guides decision-making?
- On the basis of function, what opportunities are there to redesign the product for AM? Is a hybrid solution that combines AM and traditional manufacturing methods a better option? Which AM technologies offer the best combination of characteristics for these applications?
- Do we have the right team to execute this strategy? Are we trying to do too much with too few resources? Do resource limitations indicate that we should simplify and focus on a single technology/equipment manufacturer and application?
- Would we be better served by working with a service provider or are there critical IPs to protect and learning experiences that must be gained?
- Which of these technologies should we consider bringing in-house and what are the advantages? Do we understand why we need to bring this technology in-house and the facility and supporting equipment requirements?
- Have we created a robust cost model that also accounts for technology trends and the improvements that will be available at the start of volume production or early in the ramp-up period?
- How will we capture technical risks and assure adequate resourcing to address each in turn? How quickly can we confirm these risks with internal and external customers to ensure we are working on the right problems? Is the current risk management toolset adequate for a new technology?
- How will we communicate opportunities and accomplishments with the broader organization to facilitate acceptance and support? Without compromising on the overall implementation strategy, is an early victory possible to bolster the team's confidence and organizational acceptance?
- How can we use the opportunities presented by a digital production methodology to improve the level of quality of our products?
- What is the best approach for integrating AM into our production processes? What aspects of our processes require change to achieve world-class performance?

section one

Additive manufacturing background and potential

chapter one

Additive manufacturing overview

Industrial companies have experienced challenges integrating additive manufacturing (AM) technology into their culture. Shortage of talents, lack of organizational technology deployment strategies, and a deficit of executive management technological understanding lead to a dysfunctional approach in AM deployment throughout the enterprise. Most use the term "additive manufacturing" to describe a broad group of polymer and metallic technologies. However, critical underappreciated distinctions lie in the differences among these technologies that affect how they should be utilized within companies.

Motivation and background

The motivation for writing this book originally stemmed from a conversation the authors had regarding a general lack of understanding of the deployment of AM in companies. Most conferences that highlighted AM talked only of technical achievements related to materials, machines, applications, or clever inventions. However, the authors noticed a trend evolving within the industry. Key subject matter experts in AM were often switching companies. The motivation for this recent trend in AM expertise migration seemed to be twofold.

First, AM is a newer technology and sourcing talented staff with AM expertise familiar with this technology is difficult; thereby, the law of supply and demand takes control and job candidates were embracing higher salaries being offered by other companies. However, after many personal conversations with experts in the field, things seemed more dystopian than could be explained by a mere salary increase as a cause to switch affiliations. Overwhelmingly, AM experts cite a major lack of understanding on how to effectively deploy the various technologies within industry by executive leadership and middle management. Often, various failed strategies of technology insertion would be the product of companies trying to force the technology into change resistant company cultures. Industrial leaders would attempt to replace traditionally manufactured components (machined, cast, etc.) with AM versions of the same components with no economic benefit. Expensive AM machines would be purchased by companies without clear roadmaps being developed to establish the industrialization path necessary to ultimately build parts repeatedly on the machines as required for production. Ultimately, the AM experts at various companies would see the flawed strategies and leave in search of an opportunity with more influence at a different company, only to become disappointed to see similar executive behaviors elsewhere. This dynamic cycle of AM expert migration is well known by the specialized AM talent acquisition consultants who have heard this complaint time and time again.

After pondering on this challenge, the authors agreed that what is needed is an education of management and strategic leadership within industry to appropriately develop a systematic way of deploying AM within companies. Thankfully, change management tools already exist to assist with the effort of adopting new technologies into company cultures. Using these proven tools coupled with the experience of AM practitioners, this book showcases how to effectively integrate AM industrial adoption within industry from a management perspective.

Technology introduction

There are many different ways to describe "additive manufacturing." What once began as a technology known as "rapid prototyping" in the 1990s, building successive layers of a material with a controlled shape led a recognized method of manufacturing. Along the way in the 2000s, companies began using this technology to not only make prototypes, but also as a production scaled manufacturing solution. Thus, the term "additive manufacturing" was born. ASTM F2792 released the commonly accepted definition of "additive manufacturing." The definition reads,

> It is defined as the process of joining materials to make objects from 3D model data, usually layer upon layer, as opposed to subtractive manufacturing methodologies. Synonyms are additive fabrication, additive processes, additive techniques, additive layer manufacturing, layer manufacturing, and freeform fabrication [1].

Early adopters in the field included aerospace companies looking for weight savings and a variety of healthcare companies looking for customized manufacturing solutions for medical implants, hearing aids, and dental repair.

AM represents a space that includes a number of different types of technology. There are different machines for polymers, metals, and even ceramics. Generally, AM technology is divided up into categories where each process varies in feedstock type and energy system. When first exposed to AM, it can become quite overwhelming to differentiate one machine type from another with so many machines to choose from. Next, there is the confusion on material choices. Which machines process for which specific material? Why aren't all materials available for all processes? These are typical questions managers often have when first trying to understand the space of AM. With so much confusion within the industry, why even bother with understanding the AM space? What makes AM so appealing to be used as a manufacturing solution?

Largely, there are a number of reasons why AM is valuable to industry, but the biggest reasons include (1) shortened product development lead time via the absence of tooling fabrication and quick physical prototypes; (2) design performance gains such as weight and cost reduction and multifunctionality; (3) logistical savings due to simplification of assemblies through design integration; (4) product differentiation through clever designs can yield marketing benefits; (5) customized material solutions that take advantage of inherent AM processing physics; and (6) software solutions that enhance the design, build, and verify aspects of the technology.

Given the six areas of value above, note that the true intellectual property (IP) advantage for industry in the space of AM lies in the software enhancements, design of products, and how new materials are developed for the technology. Though perhaps a controversial perspective, the authors believe that the critical IP of AM does not lie in the knowledge of processing AM parts. Of course, processing AM parts in machines is critical to the value chain. However, machines are becoming more common, feedstock is becoming more repeatable, machines are more capable, and more companies exist with knowledge on how to operate AM equipment successfully. Though once a coveted skill by companies a mere 5 years ago, due to a higher overall manufacturing readiness of the various technologies and greater competition, processing AM parts is being commoditized over time.

Now that value has been defined for the technology, let's look at how AM is categorized in terms of the technology itself. 3D Hubs published the overview shown in Figure 1.1, which illustrates the different types of AM technology that currently exist.

Vat photopolymerization is a liquid-based technology that relates exclusively to polymers and is largely used for prototypes within industry. The polymer-based material extrusion technology has been predominantly used for quick-response tooling and production assembly aids. Like Vat photopolymerization, material jetting is also used for prototypes, albeit with higher accuracy.

Binder jetting technology is an interesting technology in the field of AM. Sand casting patterns are used often in the automotive industry for prototype engine development. Metallic binder jet parts can also be used

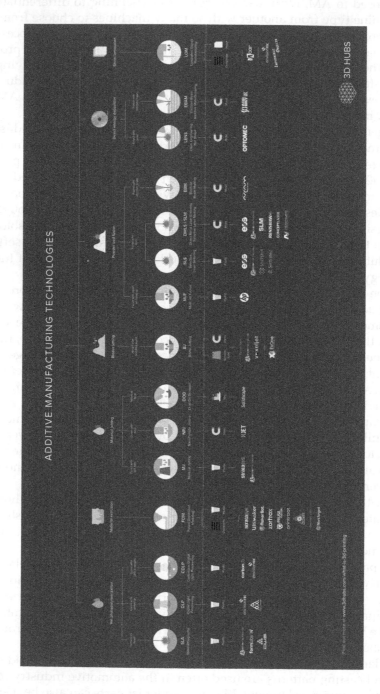

Figure 1.1 AM processes, polymer and metal. (With permission from 3D Hubs.)

as affordable end-use metal parts that do not exhibit a strong need for high mechanical integrity. Though not shown on the chart, binder jetting has also explored the space of ceramics as well.

Powder bed fusion makes up the most popular category of AM. Its popularity arises from the ability to make intricate design with high mechanical property structural integrity in both metals and polymers. This technology is broadly used in healthcare, aerospace, and automotive racing sectors. There are two energy sources in common usage today: laser and electron beam.

Direct energy deposition is a large-scale additive metals process that is often used for repair of existing large-scale parts or fabrication of preform shapes. Often used for bulk structure made from higher-value materials, such as Titanium and Nickel alloys, direct energy deposition technology has found a niche in the aerospace market.

Finally, sheet lamination from paper is an older technology for AM and was mostly used for prototyping needs. Though not adopted quite yet, sheet lamination from composites is a relatively newer technology that is gaining attention of the aerospace and automotive sectors for the ability to produce composite parts.

It is critical to comprehend that the applicability of AM goes beyond a specific technology type. Even more important to understand is that the vast majority of people on the periphery of AM are only familiar with one or two different types of AM. For example, they believe that AM is only metal laser powder bed or that AM is only fused deposition modeling (FDM) of polymer. Because most people do not understand the full spectrum of AM, it is difficult to communicate the best AM strategies or roadmaps within companies. The authors have seen various executive managers broadly make statements that apply theoretical cost savings of AM to existing product lines using (1) incorrect AM technology, (2) unrealistic production volumes, (3) parts that exceed the build volume or scale of the incorrect AM technology listed as #1, and (4) unrealistic cost savings projections based on dangerous assumptions.

For example, picture a situation where you find yourself in a boardroom full of executives and they are discussing how AM powder bed fusion technology, when applied to very large aircraft structure, will save millions of dollars of cost based on a 20% weight savings when applied on a high-rate production aircraft. Problem one, powder bed fusion technology does not have commercial machines to make massive aircraft structure in today's time frame. That would typically be a directed energy deposition AM metallic process if AM is even needed in the first place. Problem two, projecting accurate cost savings based on weight savings is never a clear situation. Problem three, no AM technology on the market today can really keep up with producing massive metallic AM structure on a high-rate production basis. As ridiculous as this example sounds to the trained AM practitioner, the authors have personally witnessed several similar discussions occurring in industry boardrooms remarkably often. It is this type of misinformation that necessitates a retraining of executives and, frankly, is the justification of this book.

Based on the previous example, you may be asking yourself, how do I know which AM technology to use for my specific application? If you find yourself asking this type of question, we should back up a bit. A better first question would be, should I even consider using AM technology at all?

In my career, I have now seen a simple constant wall thickness hollow cylinder (pipe) that was AM printed on two separate occasions at two different companies. One AM pipe was printed out of polymer, and another pipe was printed out of a nickel-based superalloy. Each engineer requesting the pipe to be AM constructed had good intentions; however, they were led astray with unbridled enthusiasm regarding AM technology. The first pipe request could have been filled by going to your local hardware store and picking up PVC pipe from the stockpile and having it cut to the appropriate length. The second pipe could have been easily sourced from an industrial tubing supply store and cut to length. The main point here is to really understand the economics driving your decision to choose AM.

Let's consider that your part may actually need to be constructed from AM technology because it is difficult to produce, very complicated, highly customized, too costly, etc. Now, which AM technology should be used? At this point, please refer back to Figure 1.1. The authors find it easier to start the filtering process with the material selection. (1) Is the part composite, polymer, ceramic, or metal? Once that is answered, the second question should be (2) does it need to be that material? Is the part currently that material due to legacy requirements that perhaps do not apply anymore? Or, is there a legitimate reason for that part to be that material type? Once answered, then you can reference the third row of Figure 1.1 to start to filtering among the different AM technology types.

The next step has two paths. (3a) If the part is determined to be some kind of tooling pattern made from wax or sand, then you are really limited to two choices: drop on demand wax patterns from the Solidscape process (predominately used in the jewelry industry) or binder jetting sandcast tooling patterns. (3b) If your part is metallic, then you have a number of choices. It is often easier at this point to filter by the complexity of part and the demand for the part to be in a structurally loaded environment. For example, of the metal process, nanoparticle jetting offers the highest complexity at the lowest structural integrity. Refer to Figure 1.2 to see the relationship of the metal processes listed in Figure 1.1 to part feature detail and structural integrity.

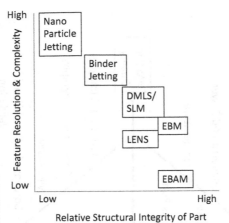

Figure 1.2 Resolution of parts relative to structural integrity.

Now, it is assumed that you will want to ensure that you will produce a high-quality part using AM in a repeatable way with reoccurring steady-rate production. But the path to reliable part quality with AM will require process-oriented leadership to establish the necessary level of process control, as many of the current generation technologies lack critical capabilities. There are many factors in the AM build process that directly contribute to part quality. Adapted from Lindemann [2], many factors that control AM powder bed part quality are shown in Figure 1.3, as an example.

We have been intentionally brief in explaining the primary AM technologies as this book is not intended to be a technical manual for any specific method or application, yet the fundamental awareness of these variables and constraints on the various AM technologies is critical to proper application within your company. As a business leader tasked with growing and preserving your company's health, a basic understanding of this broadly disruptive technology area will be critical to formulating an effective strategy. There is still time, as according to a Boston Consulting Group survey in 2016 [3], only 34% of U.S.-based manufacturing executives have even implemented an AM strategy. The reality is that many of these implementations have been one-dimensional, strategically flawed, or too limited in organizational scope to deliver the desired impact.

The field of AM both has been around for a long time and is still in its early days. The current generation of equipment can construct truly impressive hardware at useful build rates, but much more is required to justify its application in industry. The task in front of us as an industry today is industrializing AM for serial production, delivering consistently high-quality parts with differentiated customer value. In the next chapter, we will begin to explore this vast potential for AM, both opportunities and pitfalls.

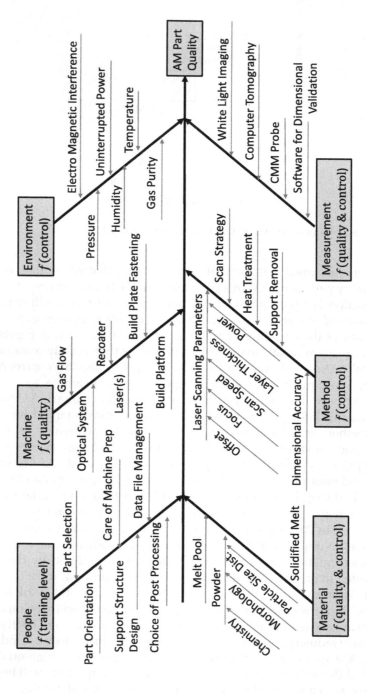

Figure 1.3 Example of factors affecting AM part quality.

References

1. ASTM F2792. https://www.astm.org/Standards/F2792.htm.
2. Lindemann, C. (2017). Systematic Approach for Cost Efficient Design and Planning with Additive Manufacturing. *Doctoral Dissertation*, University of Paderborn, Germany.
3. www.bcg.com/publications/2016/lean-manufacturing-technology-digital-sprinting-to-value-industry-40.aspx.

chapter two

Additive manufacturing potential for industry

The outlook for additive manufacturing is promising. Though skill short-ages currently exist in the field, opportunities emerge in the education sector to address these shortages. Metal additive manufacturing (AM) machines are becoming larger, more productive, and more affordable. Strong consumer customization will drive further adoption and drive the AM industry more. Currently, for powder bed fusion technologies, weight sensitivity driven through novel design will continue to drive adoption in space and satellite applications. Eventually, the software aspects of the technology will consolidate and catch up to the capability of the machines.

There is no shortage for AM use within industry. In both polymer and metals, successful applications have been demonstrated from tooling to prototyping and highly customized consumer products. It is true that AM technology is simply another tool in the toolbox of the engineer, but it is a special tool. For the engineer in industry, AM is similar to a multi-tool utility plier/knife system that every engineer has in his or her toolbox, capable of multiple functions from the same technology. If you need a prototype design to hold in your hand in a few hours, open up the AM desktop printer device feature. If you need an assembly rig designed for machining, call up the industrialized Stratasys polymer machine utility. If you need a customized artificial implant for rapid surgery, open up the electron beam melting titanium processing function. AM is broad enough to cover multiple technologies, multiple material choices and enables the very soul of multifunctional design while at the same time saving pre-cious lead time over conventional tooling.

However, diving deeper in these benefits requires a true understand-ing of the different types of AM technologies. On both the polymer and metals sides of AM, systems are growing in scale and becoming more industrialized. What once began as a machine that made prototype plas-tic parts quickly from "weed whacker" line melted through a souped-up hot glue gun mounted to a gantry has now become a machine that can produce an entire car chassis using injection-molding resin as a feedstock. On the AM metals side, powder bed fusion machines were very expen-sive, rare to find in industry, and effectively limited to an 8" cubic vol-ume of build chamber size, capable of printing roughly a pair of standard gaming dice per hour. Now, recent announcements from General Electric, Adira, and other machine manufacturers place machines in the 1 m by 1 m build volume, and printing rates continue to grow with multi-laser systems, increasingly powerful lasers, and more productive overall systems. These scales and speeds of machines were only a dream a few years ago.

So, how does this impact industrialization potential? At the rate of advancement within industry, the capability of what AM can produce in terms of value can become staggering. Recently, a Chinese company heralded the achievement of producing a single bulkhead section of the J-20 fighter aircraft using directed energy deposition, an AM type of technology. This is important from a technology development standpoint, but what about from a workforce development standpoint? This is possibly THE most challenging aspect to AM, at least within the United States.

The National Association of Manufacturers illustrates a dire picture for the state of the manufacturing industry in the United States. Despite manufacturing supporting millions of U.S. jobs, higher than average compensation, and desirable healthcare benefits, there is a stigma associated within the manufacturing industry that leads to less students enrolling in manufacturing-based engineering and trades. The U.S. manufacturing workforce is behind in higher education, not competitive in Math and Science skills and the U.S. lags significantly in graduating engineers (MAPI, [1]). With this reality in mind, the largest opportunity for growth in AM technology is workforce based.

The AM industrial community of skilled practitioners is currently in short supply. This applies to a broad group of individuals; consider the engineers fresh out of college with a single semester of AM exposure, the machine operators and technicians, and the salespeople with decent technical knowledge of AM. In addition, the industry has a massive gap in the project managers, managers, and executives seasoned in AM technology and its application. But far worse is a shortage of skilled and seasoned designers capable of understanding the design intricacies of AM. This challenge stems from decades of design for manufacturability pedagogy driven into designer's heads, leading them down paths toward nonoptimized design solutions based on legacy, self-imposed manufacturing constraints. Most of these constraints do not exist in the world of AM, yet they are still being taught and reinforced in many companies.

Despite the labor shortage challenges with AM, the industry itself has quite a bit of potential. Trends in products becoming more customized and personalized for consumers have pushed the boundaries of conventional large-scale fabrication solutions such as castings and injection molding. For example, Bentley Motors uses AM to produce metallic monograms for the owners of the vehicles to be placed within the automotive dashboard to personalize the product. Despite this example residing in the world of luxury markets, the authors believe this personalization concept will soon be the norm in the higher production volume markets. This trend will be an outcome of higher production rate designed machines and lower-priced AM machines, both of which are currently in development.

Meanwhile, trends in consumer impatience requiring quick response fabrication lead times permeate the product world. Here, with the absence of required tooling, AM supplies an answer in the form of streamlined supply chain logistics. "Distributed spare parts production becomes practical as AM machines become less capital intensive, more autonomous and offer short production cycles" [2]. With AM providing quick response solutions for out of production spare parts, another level of opportunity unfolds to showcase the true power of AM. For example, think about the automotive restoration industry for a minute. Just how large is it? According to an article on the website Second Chance Garage [3], which specialized in car restoration, Peter MacGillivray estimates the industry worth at $1.9 Billion USD. The article explains that today this generally yields small margins for the automotive manufacturing shops that supply this industry. But what if one of these small fabrication shops were to procure an AM machine and a gifted designer to model the older parts and personalize parts for the younger car restorers who, according to the article, wish to personalize everything in their lives. This trend of car hardware personalization would start to democratize this capability for car restorers. No longer would this type of work only reside at the luxury end of the scale but could soon be routine and broadly available at your local car restoration garage.

Motorsports have also utilized AM in the past, and the authors feel this trend will continue as new materials are developed and machines become more efficient and scaled in size. There are a number of presentations detailing the usage of AM by most all competitive racing teams over the past decade: from wind tunnel models using stereolithography, to quick jigs and fixtures using fused deposition modeling to lightweight functional bracketry. The automotive motorsport industry's usage of AM has become ubiquitous across the racing circuits. As recent as January 2017, McClaren Racing and Stratasys formed a formal business partnership to bring fused deposition modeling capability to production tooling and customized racing components. Moving forward, motorsports teams will continue to adopt and proliferate AM technology within their organizations to gain competitive advantages in rapid assembly, vehicle performance through grams of weight savings in components, and speedy validation of new designs.

AM continues to play a significant role in industries that are extremely weight sensitive. Satellite and space exploration sectors have traditionally been receptive to AM, and they continue to offer a consumer pull for lightweight designs of bracket structures. The NASA-sponsored Jet Propulsion Lab (JPL) has gone beyond lightweight structure to showcase

AM capabilities in a multifunctional way in a series of novel designs. JPL has now developed entire wheel systems and control instrumentation using AM as the technology to deliver true multifunctional performance to rover design. Ideally, this is a perfect insertion point for AM design technology for a number of reasons. (1) The rover is incredibly weight sensitive. NASA estimates that it costs $10,000 USD to put a pound of payload in Earth orbit. Because in properly designed AM systems every ounce saved can be significant, the $10,000 value equation allows up-front engineering design labor costs to be absorbed into the program cost easier than if a program had no distinct value equation. (2) The rover is unmanned. As such, NASA [4] does not have the same level of arduous mechanical property substantiation requirements, that the Federal Aviation Administration (FAA) levies on AM flight-certified hardware. This allows JPL to push the boundaries within the design space to truly optimize their design freedom by including AM in more load-bearing structures.

Besides the obvious weight-sensitive industries of spacecraft, aviation, and high-performance motorsports, there exist other niche opportunities for AM potential. Typically, you might find AM opportunities in any type of racing. For example, customized high-performance racing bicycles, boats, and drones. Basically, anywhere a structure is required to be lightweight for race benefits, AM may offer an opportunity space. Figure 2.1 plots our estimates of AM insertion potential by weight sensitivity.

As you can see from Figure 2.1, a few themes emerge for near-term application insertion potential of AM powder bed fusion technology – (1) either

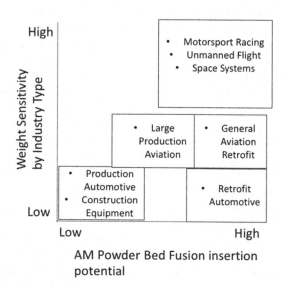

Figure 2.1 Weight sensitivity to AM insertion.

your AM design opportunity needs to be weight sensitive, such as racing; (2) low rate of production oriented such as retrofit automotive; or (3) a combination of 1 and 2, such as general aviation retrofit. It should be noted that these themes in AM application reflect the current state of the technology, which will continue to mature and progress.

AM powder bed fusion machines are becoming less expensive and processing parts at a faster rate. This increase in productivity and decrease in initial investment capital cost will create a trend that will enhance the attractiveness of existing AM potential as shown in Figure 2.2. In addition, lower equipment capital costs, faster easy-to-use machines, and more robust design software open up new areas of AM insertion potential, that higher-rate production industries did not typically consider AM may start adopting the technology. The focus will start to shift from weight-sensitive areas of adoption to more user customization as a primary beneficial pull from AM powder bed fusion technology as shown in Figure 2.3.

Given the relatively high rate of change within the AM industry, practitioners within the AM industry should expect the unexpected. The trend of faster, lower-cost AM machines and more capable design software is the current direction; however, do not completely count out entirely new AM machine technology as a potential for a major disruption in the area of AM. Significantly enhanced AM machine flexibility is not outside the realm of possibility. For example, imagine a machine that has the capability of producing multi-material (polymer and metal) structures with electronics and adhesives within the same machine. The ability to produce in a single part what is currently a complete system of components

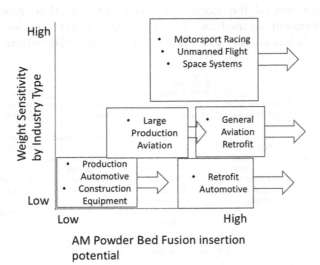

Figure 2.2 AM machines as they become faster and less expensive.

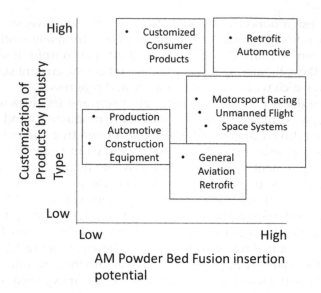

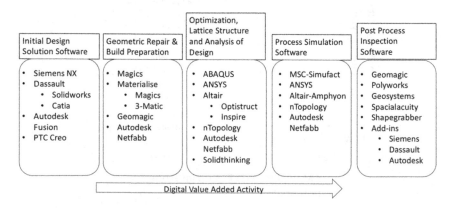

Figure 2.3 Customization growth.

represents a new level of design. This would completely shift the current trend as a step change and expand to completely new and unanticipated application spaces.

So far, we have only scratched the surface of AM potential. Similar to how ketchup and mustard usage increases with hot dog sales, AM-specific software solutions emerge seemingly out of nowhere as AM machine usage spikes around the globe. Let's look closer at this trend of software development. At the time of writing, there is a tremendous amount of software solutions for AM. Figure 2.4 indicates the current space of

Initial Design Solution Software	Geometric Repair & Build Preparation	Optimization, Lattice Structure and Analysis of Design	Process Simulation Software	Post Process Inspection Software

• Siemens NX
• Dassault
  • Solidworks
  • Catia
• Autodesk Fusion
• PTC Creo
	• Magics
	• Materialise
	  • Magics
	  • 3-Matic
	• Geomagic
	• Autodesk Netfabb
		• ABAQUS
		• ANSYS
		• Altair
		  • Optistruct
		  • Inspire
		• nTopology
		• Autodesk Netfabb
		• Solidthinking
			• MSC-Simufact
			• ANSYS
			• Altair-Amphyon
			• nTopology
			• Autodesk Netfabb
				• Geomagic
				• Polyworks
				• Geosystems
				• Spacialacuity
				• Shapegrabber
				• Add-ins
				  • Siemens
				  • Dassault
				  • Autodesk

Digital Value Added Activity →

Figure 2.4 Current AM software solution landscape.

software mapped to what function the software solution provides. Each software is at a different level of maturity with respect to its peers and let us also add a very large disclaimer to this list. We hesitate to even include a graphic of the software landscape, because by the time this book is published, this list will be obsolete. The rate of change in the AM software industry is measured in weeks and months, not years. New names will be added to this list, software names will be changed, and some software will not even exist anymore. We provide Figure 2.4 primarily to convey the absolute confusion that exists in the AM software design space today.

The good news is we are starting to see more consolidation in the field of AM software development. Only a few years ago, many different pieces of software were needed to drive value through the AM digital chain. Each software requires annual maintenance costs, technical training, and perhaps even different computing hardware requirements. If you want to quickly alienate the information technology (IT) staff within your organization, show them Figure 2.4 and tell them we need to figure out a way to integrate all of these different software packages into our IT infrastructure in order to produce a quality AM part.

Many of the AM equipment manufacturers are also working to secure this part of the AM value chain, with numerous new build preprocessors on the market in 2018. Luckily, detecting the massive potential sales opportunities, large traditional CAD companies such as Siemens, Dassault, and Autodesk are developing total system solutions for all the software functions for industry. This is a big relief to the AM industry stuck with having to pay and manage many different software tools to produce AM parts.

AM organizational maturity model

Companies naturally go through a maturation process with respect to AM. Marc Saunders developed a conceptual framework to understand how companies use AM technology and progress in their understanding. It is widely used in teaching AM content, is often seen at AM conferences in various forms, and falls close to the authors' experiences.

Let's walk through the Saunders' Renishaw Staircase model in Figure 2.5 to see how the model is applied. To understand AM, a company generally starts using AM technology to solve rapid prototyping and tooling challenges early in their adoption of the technology, which is represented by the base of the staircase. In this phase of AM maturation, a company merely becomes used to the concept of AM. The company may buy a couple of polymer 3D printers or source simple hardware from an outside vendor and begin to apply parts from that process within the factory setting. Rapid fixtures, jigs, and tooling start becoming increasingly visible application areas within the business. Once a company becomes

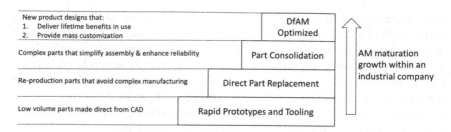

Figure 2.5 Saunders' Renishaw staircase. (With permission from Marc Saunders.)

familiar with this phase of their AM journey, they begin to branch out to the next phase, the direct part replacement phase.

In the direct part replacement phase, the company begins to explore directly replacing some of their existing part portfolio with AM parts. Often, the need to fix current supply chain issues with conventional technology drives this type of investigation. These AM part candidates are generally parts with low complexity (polymer or metal) that can be easily fabricated using conventional technologies. After one or likely more failed attempts to mine cost-saving opportunities from this effort, the company starts to learn that they must start extracting more value from AM technology. The company begins to learn how to think beyond a component level and into subassemblies to consolidate many conventionally manufactured components into a single AM redesigned solution. This awakening within the company begins the part consolidation phase of the company's understanding of AM value.

In the part consolidation phase, the company is now evaluating subassemblies and redesigning these subassemblies into an AM solution with incredible complexity. This strategy is like comparing a bushel of conventionally manufactured parts to a single AM apple to drive value. In this phase, you may start to see some return on investment in AM design in the form of cost savings. Many companies using AM technology for design are stuck in this phase. You can easily see on LinkedIn, at any given moment, pictures and case studies by companies showing part consolidation of an assembly filled with many fasteners and clunky fittings, flanges, and brackets and then a glossy highly CAD-rendered image of an AM solution concept that generally doesn't accomplish much more than the elimination of most or all of the clunky hardware. What these companies fail to understand is the next phase in AM understanding, the design for additive manufacturing (DfAM) optimization phase.

In the DfAM optimization phase, designs are now intended to be AM design solutions from the conceptual level. Meaning that AM design complexity for feature detail is now being considered upfront of design; at the very beginning of the design requirement stage. Now, at this point, a clever and skilled design engineer can use the power of AM design freedom to drive incredible value in the design. In practice, you may have seen unusual organic shapes that have been optimized for the specific load paths the part will incur during its service life. It is at this point that true AM value is being derived through the design engineering levels of the organizations. Now parts are designed with their intended life cycle and usage in mind. Walls of features not only carry loads structurally throughout the part, but may also carry fluids or integrate electrical connections, making multifunctional design a reality.

Many students of AM ask how long it takes for companies to mature through each of the phases of Saunders' Renishaw Staircase. The answer (as in all engineering related discussions) is, "it depends." It may depend on how many resources the company has to spend on AM. It would also depend on how well the company culture accepts change. It depends on the AM industry technology advancement timing as well, and we would expect that progression will accelerate in industry with the proliferation of well-publicized case studies on AM value. So, in short, it is difficult to assign specific years to each phase. But one takeaway from the staircase is that all companies reside in different phases of the model at any given point in time.

Though an incredible teaching tool to describe AM adoption and growth in companies, the authors believe that the Saunders' Renishaw Staircase is incomplete. The potential for AM optimization goes beyond the design aspect of AM. Currently, incredible amounts of research are being conducted at a number of universities, national laboratories, industry, and federal government facilities around the world in the field of AM material design.

Without going into too many technical details, high-level research is being performed on material alloy design for AM. In the past, AM metal technology has only used metals that are commonly used in traditional manufacturing technology. These metals were originally designed many decades ago for the casting or forging industries and are ubiquitous today. As a result, these materials were the first metals to be tried in AM technology – recall the direct part replacement step which would require an equivalent material to implement.

However, AM metals technology is fundamentally different from the conventional fabrication technologies in many ways. One major difference is the quick cooldown aspect of AM technologies relative to castings or forgings. Laser powder bed machines lay down small strips of molten metal that is melted by a laser. This is much faster and results in far more rapid material cooldown rates than forging or casting technology. As such, the metal materials currently being broadly used in the AM industry are suboptimized for material performance. Even the standard heat treatments are suboptimal for the same alloy composition coming out of the laser powder bed fusion process versus a legacy process. New alloys are being designed by computers, in a field of research called integrated computational materials engineering, which take advantage of the rapid cooldown physics of AM. These new alloys offer a promise of achieving structural integrity beyond what is currently offered.

Another technical potential for the AM industry is advancements in microstructure manipulation of AM material. The simplest way to describe this research is with a piece of wood, similar to a log you would throw in your fireplace. If you have ever seen a piece of wood split, you will notice an orientation in the wood that resembles the "grain" of the wood running up and down within the tree. At a microstructural level, in metal materials, a similar effect is witnessed. Research conducted at Oak Ridge National Laboratory proved that with AM, it is possible to intentionally and selectively alter the grain orientation within a single piece of metal. Back to our piece of wood example, when a tree grows, nature accomplishes this feat of intentional grain manipulation by slightly changing the grain orientation at a spot where a branch begins to grow in the tree. Nature intentionally manipulates the grains near the branches to support the extra weight of the branches and drive the load path from the branches into the tree trunk efficiently. In a similar way, and for the first time ever, AM researchers now understand how to manipulate grain textures locally within metallic structures. This is only achieved with the unique processing characteristics of an AM process.

Generally reserved for directed energy deposition (DED) metal AM systems, another unique material capability linked to AM is the ability to blend two or more material feedstock compositions together within a single AM part. This is what is known as functionally graded materials (FGM). By doing this, materials can be designed for specific function and applications. To explain this further, imagine you are a designer who wants to design a bracket. You want the bracket to be inexpensive, but you would also like the bracket to be very hard at one end of the bracket and around the holes in the bracket that see a lot of surface abrasion throughout the bracket's lifecycle. Using AM with FGM, you can simply design the part to be a cheap steel in one region of the part and a strong abrasion-resistant material, such as tungsten around the hole features and end tip of the bracket. This type of technology exists today in DED AM systems and has been demonstrated by NASA and other researchers. One day, AM systems will begin to be able to accommodate this level of material complexity. And design software will need to catch up to support this capability.

In summary, the authors see a point in the future where all the different aspects of current AM materials development trends described above will merge with the advancements in DfAM optimized phase that Saunders' Renishaw Staircase describes. The authors have coined this convergence "AM Nirvana," which is illustrated in Figure 2.6. Figure 2.7 depicts AM Nirvana as being the ultimate step of the Saunders' Renishaw Staircase. At this final stage, we will reach a point in engineering design that merges highly optimized structure design with highly optimized material design and make unique and multifunctional structures that society has never even dreamed of fabricating using conventional technologies. In the next chapter, we will take a closer look at the intellectual property trends within each AM discipline and the industry overall to further inform an effective strategy.

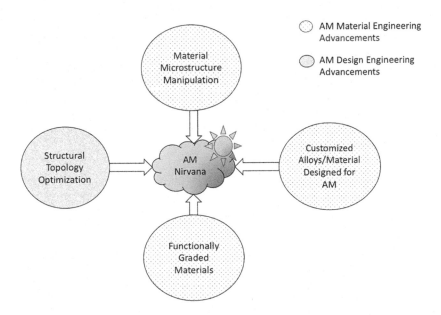

Figure 2.6 Future state of full part optimization, aka "AM Nirvana."

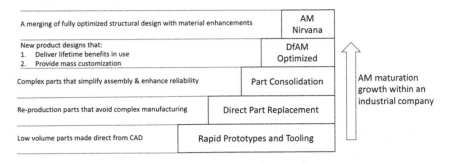

Figure 2.7 Gradual progression to "AM Nirvana."

References

1. MAPI (2012). http://www.themanufacturinginstitute.org/~/media/124212 1E7A4F45D68C2A4586540703A5/2012_Facts_About_Manufacturing_Full_ Version_High_Res.pdf
2. Khajavi, S., Partanen, J., Holmstrom, J. (January, 2014). Additive manufacturing in the spare parts supply chain. *Computers in Industry, 65*(1), 50–63.
3. www.secondchancegarage.com/public/car-restoration-industry.cfm, 2015, accessed 3/24/18.
4. www.nasa.gov/centers/marshall/news/background/facts/astp.html, accessed 5/19/18.

chapter three

Additive manufacturing intellectual property trends

Additive manufacturing (AM) is not a new field, and after an initial quiet start over the first 20 years, patent awards have increased tremendously with hundreds of new patents each year. While a decade ago, it may have been possible to personally keep track of the majority of important new patents, today there are over six new awards per day with an increasing proportion coming from outside of the United States. There is no sign of slowing down, so it is critical that business leaders establish an intellectual property plan around the technology. This will focus your team's strategy as to what aspects of the broad technology field are in versus out-of-scope for development and pursuit of patents. Some delineations are obvious such as polymer-based technologies versus metal-based ones, while others much less so including approaches to aftermarket protection and trade secret build and support strategies.

According to the United States Patent Trade Office (USPTO), AM – cooperative patent classification (CPC) code B33Y – grew by 36% in 2017 and has maintained a similar growth trend for the preceding 5 years (2013–2017). Compared with other major high-growth technologies, AM outpaces nearly all industrial sectors including even Machine Learning (G06N), Airplanes and Helicopters (B64C, which includes highly popular aerial drones), and the highly competitive Autonomous Vehicles (G05D) category. While there are a handful of other faster-growing technology areas, the publications in these spaces are limited, usually only 10–20 per year.

Figure 3.1 provides a selection of CPC classes from the very largest to the smallest and fastest growing to the most quickly shrinking. It is noteworthy that, for now, the largest class Electrical/Digital Data Processing (G06F) still generates a higher quantity of new patent awards each year than AM. Figure 3.2 illustrates a comparison between USPTO class volume and growth rate. Given visibility to pending applications, which are generally published a full 18 months after filing, combined with the restriction requirement that splits applications into separate patents for each invention, it appears likely that AM will surge to the forefront in terms of issued patents over the next few years, and possibly as early as the end of 2018.

CPC Class	Code	2017	CAGR
Air-Cushion Vehicles	B60V	14	63%
Smokers' Requisites	A24F	850	45%
Digital Computers...Computation Is Effected Mechanically	G06C	4	41%
Additive Manufacturing	B33Y	2319	36%
Computer Systems Based on Specific Computational Models	G06N	4215	34%
Systems for Controlling or Regulating Non-Electric Variables	G05D	3558	28%
Aeroplanes and Helicopters	B64C	2705	26%
Convertible Vehicles	B60F	57	22%
Casting of Metals	B22D	621	11%
Transmission	H04B	10039	8%
Automatic Musical Instruments	G10F	9	6%
Electrical Digital Data Processing	G06F	57506	5%
Micromechanical Devices	B81B	734	5%
Processes Using Microorganisms	C12R	242	3%
Semiconductor Devices	H01L	28236	1%
Climate Change Mitigation Technologies - Transportation	Y02T	5478	-7%
Edible Oils of Fats	A23D	147	-18%
Making Metal Chains	B21L	3	-24%
Analogue Computers	G06G	44	-29%

Figure 3.1 Selection of CPC classes, patent awards, and compound annual growth rate.

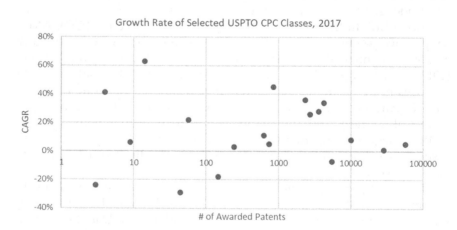

Figure 3.2 Comparison of USPTO CPC class volume and growth rate.

Large companies are gaining AM-related patents far more often than researchers at universities or individuals, with more than 80% of all issued patents going to companies. Aerospace leaders General Electric Co. and Boeing have been leading the way the past few years (generally maintaining top 25 positions in terms of overall grants), although new machine manufacturers also have a very strong rate of applications, as

Company	#
General Electric Co.	89
Xerox	78
Boeing	50
Desktop Metal	48
Hewlett-Packard	48

Figure 3.3 2017 patent applications relating to 3D printing.

shown in Figure 3.3 for 2017. While not in the top five, established 3D printing company Stratasys narrowly missed the cutoff with 40 new applications last year. Notably, the traditional 2D printing companies have made strong commitments to AM, with Hewlett-Packard (HP) and Xerox regularly making news in the field. At the 2018 International Manufacturing Technology Show, HP announced a collaboration with Volkswagen and GKN Powder Metallurgy around their new "Metal Jet" technology [1], a form of binder jetting. On display were various nonstructural components such as individualized key rings and exterior-mounted name plates, although they indicated their intent to manufacture fully safety-certified metal parts for future applications.

AM is undoubtedly one of the most powerful technology trends of the last several years and, when intelligently combined with Machine Learning and Artificial Intelligence advances over the past decade, will be a powerful transformative force in the manufacturing industry globally. Most AM processes, especially the more complex ones such as laser powder bed fusion (L-PBF), are incredibly data rich and already produce dozens of gigabytes of sensor data with each build. The ability to harness this massive data set is of critical importance to improving this unique manufacturing process and your manufacturing operational performance. It has already demonstrated the ability to enhance the metallurgical and dimensional quality and, therefore, yield of the process, as well as the equipment uptime and reliability with a data-driven preventative maintenance strategy. This trend is visible as an increase over the last decade in inventions around data acquisition and correspondingly the control and regulation of the process, which is expected to continue.

Beyond the management of this data, the ownership of said data could be an area of contention between service providers and the design owners. Companies such as Heico, which focus on providing FAA-PMA (Federal Aviation Administration – Parts Manufacturer Approval) engine parts and components, are keenly interested in the opportunity presented by modern reverse-engineering and digital-manufacturing technologies [2]. While they have not yet produced airworthy parts, noncertified hardware produced by various AM methods such as production tooling, casting cores, and inspection tools are in routine use [3]; exactly the expected progression noted in the DfAM staircase. Large original equipment manufacturers (OEMs) such as Boeing, General Electric, and Honeywell International are well aware of the looming threat posed by this technology and have taken active steps to secure their respective digital supply chains. For example, one of the authors, Dr. David Dietrich, was recently granted U.S. Patent US10065264B2 "Apparatus and method for manufacturing an anti-counterfeit three-dimensional article" for his work while at Boeing Corporation.

There are also software solutions utilizing encryption and/or blockchain technologies such as those offered by start-up companies Identify3D, Link3D, Authentise, and others, and this trend has predictably extended to the likes of Siemens and Autodesk. These enable companies to manage production of physical objects using 3D printing methods at suppliers. Users are promised the capability to secure the original digital model data, impose production rules such as the ability to limit the quantity produced, and gain full traceability to assure quality [4–6]. Aside from data security solutions, these providers are diversifying with customized tool suites for the entire AM value chain, including design, post-processing, and shop management.

A considerable growth area for metal AM will be the printing of novel alloy compositions. Only considering aluminum alloys for a moment, the L-PBF process in combination with the high thermal conductivity and reflectivity makes it difficult to control the process and produce defect-free alloys. The strongest commercially available aluminum alloy 7075 (Al–Cu–Mg–Zn system) has shown very limited processability due to hot-cracking during the process caused by the very high cooling rates. Researchers are investigating the use of novel interventions such as the addition of the element silicon [7] or nanoparticles as nucleation sites [8].

While not a bottom-up reformulation of alloy systems to achieve a desired set of physical properties, it is a viable near-term method for improving the material selection options for AM methods. There are already significant granted patents and public domain applications in this area such as Honeywell's work with a casting-grade nickel superalloy Mar-M-247 (US20160348216A1 – nickel-based superalloys and AM processes using nickel-based superalloys [9]). We expect ongoing "modified for AM" alloy patent trends to accelerate as researchers gain improved understanding of the root causes of process-related defects with these alloy systems using new sensors.

While just a couple of examples have been provided, the strong trend toward such methods to make existing alloys more weldable using an L-PBF system is clear. Beyond that, researchers worldwide are developing new alloy systems designed to specifically exploit the unique physics of the powder bed fusion (PBF) process. High-temperature alloys will be a strong patent growth area for the hot section of turbine engines and their strategic metal manufacturing supply base. To be frank, most of the supply base is still rarely printing, and when they do print, they use conventional wrought and cast materials. Difficult to weld materials, such as high-strength aluminum alloys and copper alloys, will also continue a similar growth trajectory.

Research trends

Given the rapid nature of growth in the AM market, it is impossible to predict what the fourth decade will bring. Surprises seem like a safe prediction, perhaps some new additive or hybridization of methods with ultrafast, true 3D scanning is around the corner that shows distinct advantages over current technologies. Over the next few years, some trends seem inevitable or are much clearer with converging developments and industrialization of different aspects of the equipment. We will examine these near-term trends, first for polymer-based AM technologies and then for metal-based AM.

Outlook for polymer-based AM technologies

As was outlined in Figure 1.1 per the ASTM F2792 standard, there are four primary technologies relevant for polymer-based AM: vat polymerization, material jetting, PBF, and material extrusion. None of these technologies is dominating the polymer landscape completely, with each having their relative advantages to certain applications based on material compatibility and manufacturing limitations. The fifth polymer-based technology of laminated object manufacturing of composites is an interesting, but is an emerging technology area that requires significant maturation to achieve readiness for routine production use in industry.

Selective laser sintering in a powder bed is a proven method for making high-quality parts with good mechanical properties and, accordingly, has been used in production applications across various industries. Scalability of this technology relies heavily on improvements to the laser energy system used to sinter the powdered material, with methods to either focus the energy over an area by shaping the laser output using a light valve or directly applying it over an area using an array of diode lasers a likely next step in the technology's industrialization. Recoating of the powder is also a bottleneck in processing rate increases and an area of active innovation by researchers. With respect to materials, more advanced filled powders continue to emerge such as carbon-filled polyetherketoneketone, offered by Oxford Performance Materials and Stratasys, for example.

Recently, material jetting has been gaining market share and seems likely to be a significant player in the future as the technology continues to industrialize. Since these machines can be designed to employ multiple nozzles, the technology has inherent multi-material capabilities that give it unique advantages and strong scaling potential for high volume production.

Related to material jetting, the ability to integrate sensors into smart products using direct write technology has not realized the full promise of the technology to enable and accelerate the Internet of Things. One well-publicized emerging application is the bonding of titanium alloys to carbon fiber fan blades in turbine engines.

While cold spray is not officially recognized as a true AM process, it can also be used in a similar way to repair and join dissimilar materials with a relatively high rate of deposition. A quick side note — in the current era, one good litmus test to know if a technology is considered "additive" or not is to ask if the input to the subject process is a .stl or .amf file format. Only AM technologies require .stl or .amf file formats.

In another technology area, primarily due to their low cost, material extrusion systems will be deployed in innovative ways, especially for dispensing construction materials such as concrete, food products (3D printed pizza and pasta, for example), and bioengineering applications including living tissues. Thermoplastics have been the most commonly available material for extrusion, with engineering grade materials such as ULTEM 9085 being popular in the aerospace industry for its excellent combination of temperature capability, strength, and fire-smoke-toxicity properties. Additional material variety in the future should be investigated as softer, more rubberlike materials, and composites are emerging on the market. We expect high growth in a wide variety of medical applications and continued widespread usage for applications with relatively simple geometries such as aircraft ducting and tooling. New derivations of ULTEM will continue to be developed as requirements for higher structural components in hotter environments continue to flourish. Given the Achilles heel of the fused deposition modeling (FDM) polymer process has always been the mechanical bonding between layers, we expect future systems to evolve to solve this challenge with in-situ processing enhancements or perhaps post-processing changes.

Vat polymerization has advanced significantly from the early days with a single scanning laser (stereolithography apparatus) and now uses arrays of tiny mirrors to project images to cure an entire layer simultaneously (digital light processing [DLP]), driving up productivity. A wide variety of materials and corresponding properties can be achieved via this process, giving it high potential as compelling business cases continue to be identified. Aside from plastics, it can be used to manufacture ceramic and metal materials at fine feature resolution. Challenges in this technology will be scaling to larger sizes and increasing build rates, given the relatively high system complexity. Design-usable mechanical properties are also a challenge with this type of technology. Soft, rubberlike materials

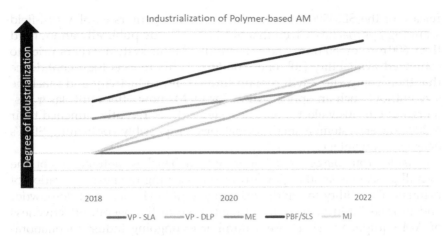

Figure 3.4 Qualitative comparison of the industrial readiness of the primary competing polymer-based AM technologies.

are already entering the market, which have improved the adoption of the technology and are even being used for car interior components. It will be interesting to see if direct manufacture of polymer matrix composites and metal matrix composites is commercialized and how such manufacturing freedom will open up new applications. Given the ability to use non-thermoplastic materials, we expect that the DLP vat polymerization method will grow in importance over the coming years. Figure 3.4 illustrates the industrialization of the various polymer-based AM technologies.

Outlook for metal-based AM technologies

As outlined in Figure 1.1 per the ASTM F2792 standard, there are three primary technologies relevant for metal-based AM: directed energy deposition (DED), PBF, and binder jetting. Similar to polymers, all of these technologies have their relative advantages to certain applications based on material compatibility and manufacturing limitations. Unlike polymers, PBF, particularly L-PBF, has clearly outpaced the other technologies for production applications. As mentioned in the polymer section, DLP vat polymerization of metals has been developing but faces significant hurdles to widespread market adoption. Similarly, material extrusion of filaments made from mixed metal powder and binders seems unlikely to pose significant threats to either PBF or DED methods. One unique opportunity for such filament-based methods may be presented by the current push for printing in zero-gravity environments.

The most visible trend in PBF has been the adoption of multi-laser systems. SLM Solutions possesses unique patents around the "stitching" of material together in overlapping scan fields, going back to the original release of the SLM500 system, while their competitors employ full-field scanning approaches. Given that SLM has already publically announced their 600 mm by 600 mm and minimum 12-laser system 2 years prior to estimated availability [10], one wonders if this reflects high confidence that their competitors will remain limited in productivity and machine size without such an option. All major OEMs now offer multi-laser systems, but curiously electro optical systems (EOS) recently launched their M300 4-laser system, a small format printer, possibly confirming SLM's near-term advantage.

Aside from bigger and better optical benches, automation of the overall process is another powerful trend shaping the next generation of printers. The ability to rapidly exchange build cylinders and de-powder and process build plates will all contribute to the overall effectiveness of AM equipment. There are a number of ongoing industry collaborations to design futuristic manufacturing cells and factories around AM. Similarly, advances in support generation and subsequent removal are reducing what is today a time and resource intensive part of the PBF process. Closely related to innovations in support geometries are advanced build process simulation that continues to improve not only the dimensional accuracy of this process, but also gives additive engineers insight into how to avoid re-coater crashes without excessively increasing support density. To inform these simulations, better sensors are needed to provide insight into the melt pool behavior, but also to provide quality assurance in a more meaningful way than current systems.

Of course, electron beams can also be used to process metal powders and have a few advantages that deftly sidestep some challenges of the laser systems. Since the processing is done at high temperature, part distortion is lower due to minimal internal stresses and also allows for processing of unique materials such as titanium aluminides. However, system improvements to remove the build cylinder to eliminate the long heating and cooling periods are needed. These systems also tend to be more costly than laser-based systems, and as multi-laser systems have grown in productivity, the overall process throughput is similar. The main downside of the technology is the partial sintering of powder, which limits feature resolution and internal geometric complexity. Innovations to address the removal and recyclability of this "cake" are going to be the key to whether EBM can scale as more than an L-PBF variant for select materials.

Aside from continuous improvements in process efficiency, safety and environmental health are big factors when working with metal powders. Next generation systems must have far more robust systems for managing powders and the generated process wastes. Equipment manufacturers are already moving from disposable filters to permanent filter designs, and further improvements should be expected given the high competition in this segment. We expect continued advances in both laser and electron beam systems to deliver greater overall cost productivity, which will expand applications and maintain PBF's position as the leading metal AM technology.

Given the potential to rapidly deposit material, DED has been employed in industry to make large, but relatively simple geometries that can be post-machined to remove the external skin. This is especially true for airframers, such as Boeing and Airbus, who value any reduction in scrap and have begun by certifying DED for non-safety critical titanium alloy aerostructures. The general technology is well understood due to a long usage history in repairs. Industry collaboration to qualify a wide variety of joints between conventional and additive material would be necessary to accelerate adoption, which is otherwise qualified for production on a case-by-case basis. Innovations to move DED capabilities closer to net shape will increase potential applications, but this alone cannot close the gap to PBF.

Hybrid computer numerically controlled machining and DED processes may be a means to alter that trajectory. This enables far more complex geometries than DED alone and is a trend worth watching, if DED is not currently in your organization's roadmap. Of even greater interest is functionally graded material enabled by DED. The ability to make robust deposits of dissimilar materials will improve system efficiency by design through the elimination of joints and/or environmentally unpalatable coating processes. Limited applications are qualified and in production today, but we project wider adoption and progress toward truly optimized structures.

Binder jetting is the wildcard of the metal technologies. Only 1 year ago, most industry experts would have said that it is mostly useful for niche applications due to material limitations and poorer mechanical properties than PBF. Steel is not widely used in aerospace, and it has some of the most demanding material requirements, which has historically limited the adoption of metal injection molding (MIM) in this sector as well. However, the automotive sector has shown strong interest in the technology through major investments (e.g., Ford and Desktop Metal). Binder jetting shares much of the promise of PBF with its fine feature resolution and elimination of tooling or fixtures, but also a more predictable path to scaling for higher volume production than even the most successful aircraft programs.

That said, significant advances in the sintering of aluminum alloys are needed to deliver on the potential. While titanium is a known moldable material, aluminum alloys are particularly challenging with both the MIM and binder jetting processes due to the low melting temperature and sintering behavior. Advances in processability of such light metals are needed before this technology displaces PBF in any industry sector. This does not seem likely in the near term, but given the heavy

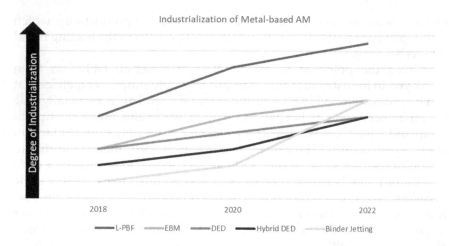

Figure 3.5 Qualitative comparison of the industrial readiness of the primary competing metal-based AM technologies.

investment, we would not be surprised to see this technology grow significantly beyond the current nonstructural applications in the coming decade. Figure 3.5 illustrates the industrial readiness of the various metal-based AM technologies.

References

1. http://h30261.www3.hp.com/~/media/Files/H/HP-IR/documents/press-release/hp-imts-announcement-9-10-18.pdf, accessed 9/21/18.
2. www.barrons.com/articles/heico-a-niche-business-with-extraordinary-returns-1498884070.
3. Joint FAA – Air Force Workshop (FAA CSTA Workshop) Qualification/Certification of Metal Additively Manufactured Parts, FAA/TC-18/3, February 2018, pp. 777–791.
4. https://identify3d.com/, accessed 9/22/18.
5. https://authentise.com/, accessed 9/22/18.
6. https://solution.link3d.co/, accessed 9/22/18.
7. Mintero-Sistiaga, M., et al. (2016). Changing the alloy composition of Al7075 for better processability by selective laser melting. *Journal of Materials Processing Technology, 238,* 437–445.
8. (2017). 3D printing of high-strength aluminium alloys. *Nature,* doi:10.1038/nature23894.
9. https://patents.google.com/patent/US20160348216A1/, accessed 9/30/18.
10. https://slm-solutions.com/investor-relations/announcements/corporate-news/slm-solutions-fy-2018-outlook-confirmed-after, accessed 9/22/18.

Creating and managing additively manufactured change

chapter four

War stories from the front line of industrializing additive manufacturing

Tremendous amounts of documented case studies exist that describe additive manufacturing (AM) saving cost, weight, and lead time in a variety of industries. However, what is conveniently left out of those glossy marketing brochures and appealing presentations are the trials, political infighting, competing priorities, organizational mismanagement, clashes with corporate culture, and the true cost of implementing AM. Given the collective experience of the authors, who have implemented AM successfully within the aerospace industry, there seem to be a number of root causes of dysfunction while implementing AM. Despite the many root causes, common themes emerge that showcase real-life challenges implementing AM.

The following content is not meant to discourage the use of AM within companies but merely to share stories that could assist any manager, project manager, director, or executive leader looking to deploy AM technology into their company and avoid common pitfalls the authors have witnessed over many years.

Brief editorial note: the following content contains real case examples of the trials, tribulations, and systemic challenges faced when implementing AM within companies. Though all examples are true, names of individuals and minor details have been altered and the specific companies are not mentioned to protect the integrity of the AM industry.

Panning for gold in Kansas

Senior leadership recently returns from a major air show and is on fire with how their competitors are reaping massive benefits from novel additive designs. Its team's efforts in this field seem to be around either tooling or incomprehensible material development science fair projects. Not to be outdone, they direct that AM immediately identify the right, high-value production applications within the business. Duly chastised, a multidisciplinary team from sourcing, manufacturing, design, structures, materials, and program management engineering convenes with a goal to comprehensively identify part candidates from the company's portfolio to be manufactured using laser powder bed fusion in production.

In the aerospace industry, there are numerous continuous improvement efforts going on at any given moment that look at changes to how parts are produced (such as changing from a cast to machined component) and procured to reduce costs. A team member experienced in this cost reduction process recommends that the AM team save time by starting with the extensive part lists already developed for this purpose to help determine where AM technology will be applied. The owners of the cost reduction initiatives are enthralled to find that there is a new technology lever to pull to secure the increasingly difficult annual cost reduction goals for their responsible areas. This AM initiative can help them out – two birds, one stone.

It does not occur to the engineering team to engage with or include key internal customers or external customer-facing colleagues in product management to identify the unmet needs that AM could solve and develop a strategic roadmap around these opportunities. Instead, the original executive goal of developing high impact applications of AM becomes co-opted by ongoing cost reduction initiatives. Despite the warnings of more knowledgeable AM personnel, the team begins mining through extensive part lists and databases to determine where the new technology will be applied. They realize that it will take their core team months to complete the full review considering all their other responsibilities.

To proceed more quickly, the team develops a simple list of questions – supposedly robust criteria (size, complexity, cost, material, annual quantity, etc.) – to ascertain the suitability of any part for AM. They then solicit help to review the entire purchased material list of many thousands of parts using this heuristic approach with respect to geometry and material. Many of these helpers are junior and have no direct AM experience, while some even lack product engineering experience. After applying the filtering checklist, the tens of thousands of parts are reduced to a more actionable list of candidates, requiring more detailed assessment by the core team. Looking at the list, a few consistent themes emerge.

Some of the candidates are immediately rejected by the core team based on awareness of subsequent processing requirements such as specialized coatings. More than 90% of the results are investment castings, mostly nickel superalloy, but some steel and die cast aluminum alloys. These castings range from compressor and turbine engine hardware, valve flow bodies, sensor housings, to heat exchanger manifolds. Despite initially discouraging standard hardware and sheet metal, the reviewers still managed to sneak a few modified fasteners and ironworker candidates through. The team realizes that the various personnel helping them process all of this data are at different skill levels in terms of AM. The brief familiarization training provided is not yielding consistent results, leading one to wonder what false negatives may still be out there.

When the AM team begins the process of developing cost estimates for these parts, they realize that some of the valve bodies and heat exchanger manifolds don't pack efficiently within – a few are even larger than – the in-house equipment, and only one or two can be manufactured at a time. Many also require a very high percentage of support material relative to part mass, sometimes exceeding 100% support by volume! Of the more reasonable and potentially addressable parts from the casting candidate pool, some have very high volumes, many thousands of units per year. Not fully appreciating the above limitations from the basic guidelines provided, some reviewers apparently passed along almost any nickel alloy casting they came across. After further filtering down the list, only three possible casting candidates are left standing.

For the non-castings, a handful of expensive forgings makes the list and initially look reasonable to the team. After talking with product owners, the team is informed that most of these are fatigue-limited designs and the AM versions would require considerable machining and potentially re-design. Upon detailed assessment, it appears this would negate most of the possible cost savings. For the few remaining strength-limited parts, it may be possible to convince the product teams to trade weight for cost on these components. However, due to the safety critical nature of many of these rotating or rotated components, there would also likely be extensive testing, both bench and engine level tests, required to substantiate the change. This is a major barrier and problematic, given that they needed to get something in production fast. The team decides to move forward with a redesign proposal for only two of the forging candidates.

Aside from the forgings, the team did find a few complex standalone machined and laser-drilled components that could conceivably be printed for a decent cost reduction and satisfy most AM principles. The undertaking has yielded less than a dozen real part candidates. The team is baffled, after initially being daunted by the prospect of hundreds of good projects from which to select and offer up to leadership. Perhaps it is possible that a few of these part-for-part conversions to AM products can generate real cost savings and return to the business. But would anyone argue that a part-based approach satisfies the original goal? A course correction from senior leadership was seriously needed, but none of them understood in advance how disappointing the results would be.

This story is not at all unique to any one company trying to develop its AM strategy, and the underlying themes captured above have been repeatedly observed by the authors at multiple aerospace companies with similar efforts heard about from other companies and industries. While reducing costs are a necessary and important goal for any company, at this time, the cases where AM can deliver significant value on a part-for-part basis are exceedingly rare. After being tasked by his leadership to personally review several thousand items, one author eventually came to refer to this flawed tactic as "panning for gold in Kansas." While some gold can be found in Kansas, it is on a recreational, rather than commercial, scale and requires meticulous gold panning methods.

Big Data Boondoggle

A variant of "panning for gold," the "Big Data Boondoggle" starts as a well-meaning effort to automate this process of identifying AM candidates. It should be better because it takes the individual judgment variability out of the equation. What is really happening here is automation of waste. In some cases, customized software may even be crafted to mine the bills of material loaded into the enterprise resource planning (ERP) system using these part selection criteria. After enormous expenses and months of effort, the disappointing results are similar to the low-tech "panning for gold" strategy. Software is unable to robustly identify system opportunities and customer pain points.

Pathfinder to nowhere

A junior design engineer named Rocky notices that the cost of the investment cast nickel superalloy Linkage he is working on looks extreme, more than three times a similar legacy part. He wonders if AM technology – Rocky became interested in AM during college where he printed a few plastic items for a senior project – can help reduce the cost of this component somehow and suggests it to his manager Scott. Having recently received word that his team lost yet another bid for a Linkage assembly program, Scott has been frustrated that the costs seem so much higher than he remembers them being when he was the one designing components. Recalling an e-mail announcing that one of the other facilities recently acquired a metal 3D printing machine, he makes a few phone calls and finally gets connected to the right team.

During the initial discussion, Rocky and Scott explain how component costs are out of control and Sourcing does not seem to be doing anything about it. They are assured by the newly assigned AM project engineer Tony that the printed material cost model shows an easy 40% cost reduction for the Linkage. Tony knows he needs to move quickly with this opportunity and demonstrate the value of AM to everyone. He doesn't know much about Linkage assemblies, but as an experienced project engineer, he realizes that the nonrecurring engineering costs of a substantial design overhaul for even a relatively simple aerospace component could be a major barrier to this project. The easiest solution will be to minimize the nature of the change and keep the switching costs to a manageable level, thus securing the cost reduction with quick payback to the business based on earlier implementation timing. Scott likes the sound of that as he's been feeling the pressure for out of control product costs from his leadership. Thus, $60,000 in project funding and Rocky's support are secured to convert the Linkage into an additive version for production use – the first non-tooling AM project for the company!

During the project kickoff, dramatic opportunities for design improvement of the entire system in which the Linkage operates are suggested by expert AM engineers. With some intensive effort, these methods look like they could simultaneously reduce system weight by more than 30% and system cost by at least 70% through integration of other components, elimination of most standard hardware to join them, and the associated inspection labor. These proposals are unheeded by Tony and Rocky who are accountable to deliver the quick cost reduction goal to their respective closely monitoring leadership chains (their annual goals and bonus are even tied to this project's success). In spite of the incredible opportunity, these Design for additive manufacturing (DfAM) proposals are out of scope relative to what he and Scott already agreed and the cost model already showed that just printing it nearly as-is to the casting model will provide significant savings.

He reaffirms that the team is permitted to only make minor changes to the cast baseline such that the resulting Linkage design is virtually indistinguishable from its origin – in fact, it could be investment cast as well. The main structure remains a constant wall thickness section, the attachment points to adjacent components of the assembly are unaltered from the baseline (as this would increase the project scope), the weight and stiffness are kept identical to avoid changing any system dynamics, and so on.

Fast forward 4 months during which more than a dozen failed hardware builds (scrap) are made while trying different support strategies to achieve the required dimensions and associated tolerances for this Linkage design. Surprisingly to the team, manufacturing to an engineering drawing with complex profiles and tolerances is not quite as easy as making prototypes and simple tooling with limited requirements. Even worse, new requirements seem to be discovered every week by the ever-growing team assigned to help and they have had to make a couple of design changes along the way. The materials engineer for the Linkage project needs hundreds of mechanical test bars to validate the properties for the detailed structural analysis that no one thought was required.

These are manufactured and tested over the next 5 months, and component-specific substantiation test requirements such as vibration and actuation performance are also successfully completed. Numerous project meetings and design reviews have been held, and the internal stakeholders have agreed that this new component is acceptable for use in the intended application. Many months, thousands of hours, and more than $250,000 have been invested to bring the project to this point. It is time for Tony to take that initial, rough cost estimate and based on what the team has learned over the course of the project, refine it to take credit for the cost savings to the business. Surely, the payback will be a bit slower than promised due to unplanned nonreoccurring expense NRE, but it's still a winner.

But what's this? The cost of the baseline casting is now less than half what it was when the team started the effort! Without notifying the team, the Sourcing department already concluded negotiations of a new long-term agreement 2 months ago with the casting supplier and had engineering loosen an unnecessarily tight tolerance on a feature which allows the part to be manufactured with high yield like the legacy designs – the root cause of the "should cost" discrepancy Scott felt. Suddenly the promised 40% cost reduction is actually an 80% cost increase. The rationalization, blaming, and excuses begin, but Tony cleverly suggests that the team can market the project as a "pathfinder" to justify the expenditures. Senior Leadership is reassured that the promised value will be realized on the *next* project now that all the processes, required documentation, and system challenges have been thoroughly mapped out. Unable to lose face on their first effort, the team goes forward with an initial "production" run of 8 for usage in the field, while quietly switching back to castings for serial production. Embarrassed, Scott chastises Rocky to not waste any more time with AM and won't even answer the AM experts' requests to reconsider the original DfAM proposals.

The path that AM technology will take in your company critically depends on what problems you apply it to and how it can solve it in a unique way that delivers value to your customers. Even if this Linkage project had worked out in the end and delivered the committed cost reduction, so what? There is a forever-lost opportunity to energize your company's engineering and business teams with transformative application of DfAM to a system, rather than a part that doesn't truly impact the end customer. Instead of this laying out a brightly lit path that ignites the groundswell of innovation within your company, the AM team now find themselves in a poor position to drive change now that word has spread within the company about this failure.

Eventually, all engineered products converge toward being treated as a commodity, and we believe AM is a powerful tool for disrupting this trend by de-commoditizing your products and maintaining differentiation in the market. As we emphasized in Section I, reimagining the architecture and designing in greater functionality at a system level is where true AM value lies. As leaders, we often say we want our teams to be bold and take risks, and yet fail to truly enable these desired behaviors by shackling new thinking to counterproductive methods and misaligned goals. Ultimately, AM needs to solve a problem to be broadly adopted within your organization and the first project your team identifies must deliver on this promise, not exclusively be a "cost-take-out" opportunity for supply chain management.

Cultural resistance to change – crab mentality

One massive barrier facing the AM industry is the cultural adaptation a company goes through to adopt a newer technology such as AM. Executives and strategic thinkers often massively underestimate the amount of organizational disruption that must occur to pave a path for AM to be adopted. Often, if the leaders did recognize the amount of change necessary to clear obstacles for AM, they remain stagnant until their specific leadership calls for drastic changes. In other words, a crisis within a company must be created to receive change.

One story that highlights this point of stagnation is a story dating from 2015. An executive, named Harold Mavis, read a competitors' article in Aviation Week magazine on how the usage of AM design afforded the competitor the ability to drastically reduce weight and save costs. Not truly understanding how this was achieved by the competitor, nor the over hyped and heavily marketed benefits of the article, Harold immediately ordered his team of engineers to create an entire aerospace product made from AM.

Upon issuing the command to his sub-tier management, they immediately found engineering bodies available to immediately start work on the project. Due to the fact that the best engineers are constantly busy, the most readily available engineering talent within companies, are not the highest caliber innovators. Not recognizing this shortfall, Harold gave a directive to this team of highly available "talent" to create a complete AM product. Coincidently, this design effort was given to Alicia Redeaux, a seasoned structural engineer that had been around the company for nearly three decades to lead as the Project Manager. Alicia had been known for being difficult to work with and resistant to any change. If you want the least amount of innovation in concept design, give the project design authority to a structural engineer. Structural engineers are responsible for the structural integrity of the aircraft frame, engine, brake system, etc. Due to the immense amount of responsibility and power they hold within the teams of engineers, they are typically the most conservative natured of all aerospace engineers, therefore, the least innovative. There may be many structural engineers offended by the previous statement but based on many decades of combined aerospace of the authors, we have concluded it is almost a natural law at this point.

Back to the story, structural engineer Alicia's given goal from Harold was to make the entire aerospace product out of AM. Alicia did not engage AM expertise within the company, but rather sourced readily available design talent to design the product structure her way. After 6 months of design effort, she was instructed to at least consult with the AM experts within the company. So, after 6 months of work, Alicia held a briefing to show the product concept. Upon the unveiling, the product looked exactly like all the conventionally manufactured products with zero design innovation. A measly 10% of the product concept was to be produced via AM. When confronted about the lack of AM on the product by the AM experts, Alicia said that she didn't trust the mechanical property data from AM materials as it was "new" and "unproven", so she determined the appropriate design path included largely conventional fabrication methods. Upon showing the product design to Harold, the program was immediately cancelled and approximately $450,000 of labor and materials and 6 months had been wasted.

The previous case example is meant to signify a common challenge in the AM community that has been repeated over and over again within companies trying to implement AM. The wrong people, uneducated about AM design tactics, are put in charge of design innovation often fail in implementing AM. Even when the right people are put in charge of leading innovative efforts, all new product designs are completed as a team. As such, each team consists of different perspectives all working toward individual goals and performance metrics. Within the aerospace industry, engineering's inhibition to change is ubiquitous among all leading companies. More data can always be requested, more testing conducted, and alternative ideas questioned. This type of mentality leads to a gridlock of activity moving AM forward.

What is commonly known as "the crab mentality" [1] is a metaphor based on the behavior of a bucket full of crabs. Some of the crabs at or near the top of the bucket could escape by climbing on top of the others. However, other crabs within the bucket pull them back down to prevent any from escaping the bucket, thereby guaranteeing the entire group's demise. Within social teams, humans can behave in a similar fashion as the crabs. Often, in response to any individual team member that achieves success, others within the team will halt their progress out of envy, spite, conspiracy, or competitive feelings [2].

The authors have personally witnessed this crab mentality type of behavior when trying to implement a new technology, such as AM. As an additional example, Derrick Kern was largely inept in his job role of technology manager for AM. His experience and background had been in the traditional fabrication process of machining technology. Though very few breakthroughs have occurred in the area of machining in the past several decades, Derrick neither had the initiative nor ability to learn a new technology. He typically saw AM as a threat to his beloved machining technology.

Despite not having any formal education in management, Derrick's unquestionable amount of time spent appeasing his management over the years paid off and management placed Derrick in charge of the newly formed AM team and the machining team. Instead of learning about AM and investigating how AM and machining complement each other, Derrick saw AM as a direct threat to internal funding to his precious machining team. Like the crabs in the bucket, he worked very hard to discredit AM technology, the individuals working on AM, and by setting incredibly unrealistic goals in an effort to eliminate funding of the AM team. The lesson learned from Derrick is that management should think carefully when setting up an organizational hierarchy with respect to a newer technology and not create technological competition among teams. Typically, match a newer technology, like AM, with younger energetic leaders that have a passion for any new technology and leave conventional technology to seasoned leaders interested in maintaining the status quo.

Who's in charge here?

Whether it is due to legacy company culture or how the company was originally organized, ownership identity of AM within large-scale industrialized companies seems to be a major issue. Within the aerospace community, functions are often siloed into separate departments that maintain their individual mandates. Often, these companies are highly matrixed organizationally and niched areas of specialties emerge over time. Remember that AM is strongly cross-functional, which means no matter if the company's strategy is to outsource or insource the technology, AM touches many different functional departments.

First, the Design Engineering Department develops the AM designs. But who develops the materials to be used in AM? That answer would be the Materials and Process Department. Second, someone needs to develop the new research for AM, correct? That responsibility could fall under an Advanced Technology Department. Who would be responsible for procuring the parts from external suppliers? This would be the Supply Chain Management Department with assistance from the Manufacturing Engineering Department. If you decided to bring the technology in house, which department above would have ownership of the technology? Often, unless the executives in charge of the company are experienced with AM, this proxy war is never truly sorted out. Making matters worse, with AM receiving so much attention (and with that, resulting large budget dollars), departments could perpetually fight for ownership and strategic direction.

This poses a question; will your company be one who recognizes AM's cross-functional impact and coalesces around the technology and possibly reorganizes into integrated product teams that support AM projects. Or conversely, will your company be lost in a constant state of conflict among siloed departments with AM being pulled apart at the seams creating chaos, communication challenges, and confusion?

Engineering rigor mortis

While trying to implement AM, teams will go through Tuckman's [3] Forming, Storming, Norming, and Performing phases. If you are not familiar with these phases, I will paraphrase these phases quickly. In the "Forming" phase, a team is brought together to solve a problem. During this time, most of the team behaves in a positive, congenial fashion toward one another; feeling out each other's behavior and responsibilities. In the "Storming" phase, the teams war with each other to establish dominance and exhibit different working styles. During this phase, quite a bit of stress is endured by the team members and often there is a loss of control during meetings, and communication is strained. Next, in the "Norming" phase, teams start to get along with each other better and most accept the roles and responsibilities of others and progress is made toward the team goals. In the "Performing" phase, the actual work is completed, team members are developed, and tasks are executed well.

Though Tuckman's model above is a well-known model for teaming within companies, it seems though newer technology, such as AM, leads teams to be stuck in the "Storming" phase for longer than anticipated. The authors believe that when introducing something new and novel, engineers especially, tend to fight the new technology and provide resistance. Of course, there may be good reasons for the technology resistance; however, skepticism tells us that often it is easier to say "no" to new technology than "yes."

Incredulously speaking, there are many tools in the engineering toolbox to fight against new technology. From an engineering perspective, more data can always be requested because often engineers do not believe engineering data generated by others. As such, projects can become stuck in what we call "Engineering Rigor Mortis," or the death of a project due to over defining data requirements, expanding unrealistic testing, paralysis by analysis, and just stacking theoretical risks onto more theoretical risks, thereby driving up more testing. For AM, this type of behavior typically comes from more seasoned engineers and the conservative structural engineering community.

For example, this author has personally witnessed a 77-slide presentation prepared that detailed a thermal analysis for a new AM design. It took 6 weeks to prepare the presentation, and 95% of the presentation was completely irrelevant to the new and novel aspects of the engineering design. It was later revealed that the manager, Noel, needed a charge number to provide to his team to give them something to do because they were low on work. The team took the charge number, which by this point was acting as a jobs program, and racked up many hours of labor of analysis. Of course, the charging didn't stop there, the structures team then needed to further justify their efforts, by documenting the wasteful charging into a

77-slide presentation format, therefore, creating the ultimate Engineering Rigor Mortis.

Of course, data scope creep isn't the only tool in the engineering box of dysfunction. Cost justification can always be used as well to prohibit the adoption of new technology. For example, how many times have you heard, "we would love to implement AM in this area, but the testing costs along won't pay back the anticipated savings." Always expecting immediate payback for initial AM projects is a shortsighted mentality to have because it may only take one project to absorb the upfront non-reoccurring testing costs. After the costs for the first project are sunk, subsequent projects can reap the testing data benefits of that first project. The authors believe that oftentimes, financially justifying every action that must occur within business is a powerful way to stymie innovation.

Suckers for sunk costs

In the world of AM industrialization, it is truly interesting how different companies react to their competition. From 2011 to 2017, it seemed that every aerospace company had to publicly announce their investment in AM technology via AM machine purchases, establishing research and development centers around AM, acquisitions and mergers with powder companies, machine developers, part suppliers, software, etc. As such, this created a frenzy of activity and discussion within aerospace companies. In boardrooms all over the world, you would hear phrases such as "did you hear X - AM company bought Y- AM company?" "Our competitor X is buying an AM machine company" "Our other competitor just bought our AM supplier, now how are we to obtain parts supply?" This type of panic created a fictionalized need for companies to "move fast" in the space of AM.

But you may ask yourself, "moving fast in companies is good, right?". I would answer "yes, if you truly understand the technology you want to move fast in." That was largely the problem with companies during this time frame. Some companies would respond with the AM marketing hysteria with cautious optimism and dabble in what AM technology could do. Some companies would respond by discounting the AM technology as another fad and ignore it. Other companies would have a knee jerk reaction by acquiring large amounts of machines, plugging them in, tell everyone in the company that they are going to vertically integrate it and then ask, "where is my ROI?", naïve to the cultural mind shift and the downstream investments the company needs to go through to learn AM technology.

The authors have personally witnessed different companies behave in the three patterns above. The last example typifies several companies that we will focus on and categorically place them in the "suckers for sunk costs" theme. By purchasing large amounts and different companies' AM technology, let's look at the downstream effects of such a strategy.

1. Buying many machines, you now have to support maintenance agreements and software licenses of several different machine types. With each piece of AM equipment having substantial annual maintenance costs (of around $35,000 to $70,000), this often unanticipated cost can add up very quickly when multiple machines are acquired or leased.
2. You now need many qualified technicians to run machines. Experts have stated that the technician-to-machine ratio in the AM industry is around 3:1. That is, if you have a technician who is highly skilled and cross-trained to run multiple types of AM equipment. This skill set is extremely hard to come by. Sourcing these people is expensive and time consuming.
3. You need to have a facility to accommodate all these machines. What is not shown in the glossy marketing literature of the machine vendors is the special health and safety requirements that your company will need to endorse prior to ever even turning the machine on. This, in turn, will no doubt be costlier than you anticipate. For example, if your company wants to use multiple materials, typically machines that run aluminum and titanium need to be physically segregated from machines that run steel, nickel, etc. This is due to explosion prevention and material contamination. Get ready to install all pneumatic cleaning stations and vacuum systems to eliminate any sparking that may occur in the Al and Ti room. What about Argon evacuation gas system in case of an Argon flooding condition? Yes, another cost adder that you didn't anticipate. These examples are only a partial list of special accommodations required for an AM facility.
4. Accounting systems: Does your company have an ERP system that calculates costs accurately with the nuances of AM? How will you calculate the true cost of AM metal powder reuse? It may seem trivial, but the authors have witnessed a lot of angst aligning a system such as SAP to accurately account for powder reuse and replenishment.
5. Metallurgy: Speaking of powder reuse, will your company invest in the large amount of mechanical property testing necessary to claim powder reuse is acceptable? This is another added cost that many companies don't anticipate when setting up their own AM production facility. What about experimental testing to optimize post process heat treatment and hot isostatic pressing for destructive test coupons ... another added expense.
6. Design engineering: Will your current CAD system be able to design complicated geometry easily? Most likely not. You will need

additional software licenses beyond your traditional CAD system. This will include ancillary software that fixes the digital files created by the CAD system and transmits the part build information to the AM machine (~$35k USD per seat annually). In addition, pre-distortion modeling software that predicts the amount of distortion your part will incur during AM fabrication will most likely be needed (~$75k per seat annually). How about scanning equipment and software that compares the original CAD design to what was actually produced to see how far off dimensionally your part will be (~$10k USD per seat annually)? Despite marketing literature saying the contrary, your AM metal laser powder bed parts will not meet your original design tolerance definition after they are printed and heat treated. Either the design engineering team will need to design around the tolerance capabilities of the new equipment, or you will need to invest in the above to meet geometric intent.

By any stretch of the imagination, the above is not a complete list of changes required to vertically integrate AM technology; the above investments reflect real actions that must take place organizationally that can have significant financial ramifications that are often not considered up front in boardrooms to invest in AM technology. By this point, a reasonable executive or manager must wonder, should we really vertically integrate this technology? If so, how broad or narrow should our scope of work internally entail? Outsource only AM fab and working engineering, heat treating and post processing internally? Invest in AM equipment and run the entire operation internally? Outsource all fab and keep engineering design internal? These are strategic questions that should be determined up front prior to machine investment.

Relentless drive of organizational dysfunction

The authors are unsure of the origin of the following humorous story, but it is a story that has lasted generations. Paraphrasing the story badly to quickly get the point across, it goes something like this. An incoming manager, named Jim, who is replacing a recently fired outgoing manager, named Tom, finds his way into work for the first day on the job. On his way out, Tom, the outgoing managers provides Jim three envelopes labeled #1, #2, and #3, and states, "when things get tough in this job role and you run into problems you can't solve, unseal the letters sequentially one at a time."

The first few months on the job go quite smoothly for Jim. He is new after all and not expected to make a massive impact at first. After a few more weeks, Jim starts taking heat for late shipments, cost of poor quality, design escapes, and ballooning overhead costs. Jim reaches in his desk and opens envelope #1, the message reads "blame your predecessor." Jim holds a large executive meeting and thoughtfully placed the blame of all of the recent challenges on Tom, his predecessor. Things became quite again for a while.

About a year later, the company was again experiencing some substantial challenges in the late shipments, poor quality, overhead issues, etc. Remembering the letters, Jim opened up his desk and opened envelope #2. The letter stated, "Reorganize." Jim convinced his leaders that a reorganization was in order to resolve the challenges. After careful planning, human resource engagement, many departments were restructured over a period of several more months. The situation seemed to improve for a few more quarters.

However, once again, the company began a downward spiral of cost overruns, poor quality, and late deliveries. Jim, now a firm believer in the envelope messaging, opened the #3 envelope. It read, "prepare 3 envelopes."

Though a very funny story, this story hits a little too close to home for many readers of this book. If you have spent any time at all in industry, you are too familiar with constant restructuring initiatives meant to drive productivity and slash overhead costs within organizations. Over the course of a decade, the authors have personally witnessed functionally siloed organizations change to more horizontally structured organizations, then remain for a while. Once everyone gets used to the horizontally structured organization, then new management arrives. Naturally, the new leadership wants to leave a mark for their reputational legacy and moves the organization back to more siloed organizations. This pattern repeats within companies over and over again.

The above approach is merely the executive leadership equivalent to the Hawthorne studies. If you are not familiar with the Hawthorne studies, they were a series of experiments developed by psychologists Mayo and Roethlisberger that were performed on workers at the Hawthorne plant of Western Electric Company in the 1920s. Originally designed to study the effects of worker performance with increased electrical lighting, the study results surprised everyone. The psychologists lowered the lighting, then managers would walk out to the production line and measure the efficiency results. The psychologists then increased the lighting and the managers would visit the production line and measure the results. The results of the study indicated that increasing the lighting had nothing to do with efficiency. In fact, workers were highly responsive to the additional attention from their managers and the feeling that their managers cared about them. The Hawthorne studies indicated worker's performance is a function of job satisfaction and social issues more than specific isolated changes. Similar to the three-envelope story, often times, workers just want to know that their work matters.

1. So, what takeaways can be garnered from the above stories for managers facing AM implementation within their respective companies? How can we learn to adapt and overcome these types of situations within organizational change? Recognize organizational structures will always be dynamic within companies. Sometimes organizational restructuring can work in favor of AM by providing more organic communication, more budget, more resources, etc. However, a restructuring can also have large negative impacts as well. This could come in the form of relocation of physical AM machines, slashing of AM R&D budgets, strategic direction shifts based on core production becoming more of a priority. The authors have personally witnessed all of the above depending on the financial performance of the company at a specific point in time.

2. Understand that leadership will also change over time. If you only have one executive backer of the technology within the company and that person leaves the company, your hard-earned AM budget that you have built credibility with over the years will be harvested faster than you can imagine. Politically speaking, at a minimum, try to have more than one executive advocate for AM technology.

3. The people on the front lines of AM will affect the implementation and adoption of the technology the most. These are the technicians running the machines, the manufacturing engineers setting up builds, the design engineers running topology optimization routines on a design, and the supplier managers traveling to AM suppliers. If the people above aren't motivated to drive AM technology, no organizational change effort will yield positive benefits for AM.

Knives out

After many months politically battling different strategies, executive leadership comes to a consensus on how to move forward with AM technology. The company decides that AM technology has a fairly low technology readiness level for the products that the company makes. They determine the best approach to move forward with AM is to set up a laboratory with different AM technologies under one roof to perform research on and learn more about them. Once the technology is properly assessed and of sufficient maturity, the strategy will be to push the lessons learned in the form of specifications to the company's current supply chain. Many companies in the AM field have taken this strategy and it isn't necessarily a bad one. However, as seen below, there can be unintended consequences to this strategy.

Management determines that the research department will own one of each type of AM technology. The company sells both polymer and metal products so they match the AM machines to polymer AM and metallic AM. As such, they purchase one fused deposition modeling (FDM) machine, one selective laser sintering (SLS) polymer machine and one metal powder bed (electron-beam melting [EBM]) machine and associated quality inspection equipment such as a structured light scanner. Next, management allocates one budget to be spread across all the AM technologies to research the machines' individual limitations. Subsequently, the company hires subject matter experts for each of the respective technologies. Elizabeth Bathory was hired for SLS, Mayes hired for FDM, and Drake hired for EBM. Finally, Jayjee self-identifies as the omnipresent force behind quality inspection of AM.

Keep in mind that the company incentivizes the individuals for the performance relative to their technology with annual performance reviews, patents and how well the technology performs. Knowing that there is only a finite amount of budget to be allocated across the technologies, this sets up an internal war for resources. After the previous manager of the team left the company, Russet Burbank, the new manager of the AM team, is overwhelmed with political infighting and does little to change the behavior of the team. Despite having access to budget and after producing little technical achievement on the EBM side, Drake basically gives up and escapes the group by taking an unrelated management role elsewhere in the company.

Because of Russet's lack of effort curtailing the fighting and AM's growing popularity, others are hired into the group and immediately take sides of the brewing civil war between FDM and SLS. Chet, Dr. Jay, and Rick Herr are brought in to assist with SLS and side immediately with Elizabeth. Chet even endorses Elizabeth for an AM industry award.

Behind the scenes, Chet and Elizabeth are downplaying FDM's usefulness to internal customers. Dr. Jay and Rick Herr also do the same. The motivation for this assault on Mayes' credibility is to drive FDM's allocated budget down while increasing SLS budget. Elizabeth and Chet marginalize the effort and FDM technology that Mayes is working on by systematically pointing out flaws in FDM relative to her SLS technology. However, Mayes is a fighter, and he establishes a strong funding line for FDM with an internal customer outside of the team. Elizabeth and Chet are enraged at Mayes' funded activity and engage in mechanical property data manipulation to make SLS look attractive, outrageous technology costs are hidden regarding SLS, and untrue accusations are being made at Mayes and Jayjee, who has sided with Mayes. Meanwhile, because all the machines and people (with the exception of the virtual employee Elizabeth) are located in one room, shouting matches begin to explode.

One by one, the caustic not-a-team self-dismantles over time. This shrinking wasn't due to AM technology, but rather the lack of supervision, unethical behaviors, abusive personalities, and unintended consequences of management structuring. How could this situation have been avoided? What lessons could we learn from this case study?

1. Think carefully how individuals within AM work teams are incentivized. Consider setting group goals that force competing teams to work together and that challenge the team to think creatively.
2. Place strong management in charge of AM work teams. Management that listens and acts to employee feedback and fights to find the budget necessary for the team. Management that pays close attention to how the team interacts. Punish inappropriate behavior and be willing to remove troublemakers from the group.
3. Evaluate how AM groups will receive funding. Will they be funded internally only, through government contracts, through product teams? Will there be a competition among AM technologies for funding?
4. Evaluate talent carefully before bringing them into AM work teams. AM work is cross-functional and, therefore, requires dynamic personalities that can work cooperatively with others.

Innovate NOW!

"Managing a firm's knowledge assets is as important as managing its finances" [4]. Many companies try to innovate through "ideation sessions" meant to drive AM ideas on products … don't get me started on this nonsense. If you have ever been unlucky enough to be part of these in your career, you will know exactly what is wrong with this approach. For those unfamiliar with these workshop meetings, the idea behind innovation sessions is to gather about two dozen of your highest-paid engineering "thinkers" in a room for one day or more.

During the ideation event, the group of twenty-four engineers is subdivided into a number of sub-teams of participants. After about 2 h of listening to leadership discuss the importance of the event, teams are given a set of vague questions on how can we make our products better to stimulate thought. Next, the teams are given about two to three hours to come up with many ideas. Subsequently, the ideas are floated among the larger group and participants throw rocks (figuratively) at the ideas to eliminate the bad ideas from the good ones for another 3 h. Finally, the down selected ideas are forwarded to the executive that sponsored the event.

This ideation session approach generally amounts to $34k of wasted effort, (8 h × 24 participants × $175/h) plus the opportunity cost of what those highly paid thinkers didn't get accomplished during this time. I have been involved in over six of these events over a span of about 5 years, and none of the ideas have actually ever resulted into a funded project, moreover, a profitable new product. So, why doesn't the approach work?

First, at no point in any of the ideation sessions that I've attended were customers brought in to share their current challenges with the company's products. I'm unsure as to why this has been the case. Is it too difficult logistically to bring in a customer or hold a workshop at a customer's location? Is it an organizational structure cultural barrier that prevents engineers from speaking with customers directly? Is the company afraid of the external appearance that the company is not omnipotent?

Notwithstanding the reason why, without the customer's input in the innovation sessions, this leads to a lot of guessing of problems the customers may or may not be facing, which then leads to solutions that may or may not be relevant. Technology, like AM, is force fitted into solutions to problems that never existed in the customer's eyes.

Second, is a conference room miles away from a product line a great location for this type of event? After spending almost two decades of my life in conference rooms over my career, without a doubt, I loathe them. Cold coffee, stale doughnuts, untoasted bagels, people occupying the room's attention for no reason other than to hear themselves speak, white boards with permanent marker marks that can't be erased, erasable markers that don't function, PowerPoint presentations that don't link to the projector, distance attendees unable to hear the room discussion … just the mere thought of this environment kills innovation. This is the wrong approach. Innovative product design thinking largely occurs at the point of use. The point of use would be how/where the customer is using the product. Some of my best AM design solutions over the years came from directly observing how the customer installs and uses the product.

Third, can innovation be directed in a formal format or does it flow organically between casual conversations between engineers, customers, marketing professionals, etc.? It isn't a coincidence that many great AM design solutions are sketched out on the back of a napkin at a restaurant located at a conference with industry peers.

Fourth, do you really want the highest-paid engineering 'thinker' participants coming up with innovative ideas? Most likely not. You will also want to engage in new hires, interns, and people from other business functions that have little knowledge of the product line discussing ideas. Often, people too close to the product will have a difficult time creating new concepts for the design. On the other hand, people completely unaware of the product line will be more poised to create more creative product solutions. Obviously, you will need the seasoned engineers to stymie some of the outlandish ideas, but you may only need a few highly paid engineers in the mix of participants.

So, why do companies engage in wasteful ideation sessions? What is their motivation? In my experience, it is derived from the highest level executive requesting a strategy of more growth needed for the company. This vague strategic direction is taken by some to mean more new products must be introduced. Next, there is a misconception that all new product ideas come from engineers (when they also can come from sales, marketing, business development). Of course, then the actionable next step must logically be to lock all the highest-paid engineers in a conference room and tell them to innovate with inadequate background information and guidance.

How can we depart from the ideation session framework? For one, let's stop formally prescribing an outcome of innovation in ideation sessions. Let's recognize that innovation can occur more frequently in an informal setting. The authors suggest deploying this strategy in the form of professional goal development at the individual employee level and make sure the employee (new hires and seasoned engineers) attend several AM industry conferences that incubate ideas from a large amount of diverse thought. Also, consider having the employee travel with the sales and marketing teams to interact face to face with customers who use the product on a daily basis. This approach will no doubt lead to more innovative design thinking that may or may not require AM solutions. Remember that AM may not be the solution to new product design. It is a tool in the engineering toolbox, not the entire toolbox. Instead, AM should be regarded as a disruptive technology, not displacement technology.

To summarize, please consider the lessons learned from the previous stories to provide guidance when moving toward your own path for AM deployment. Table 4.1 summarizes the lessons learned from each story for reference.

Table 4.1 War stories summary

Story	Lesson learned
Panning for gold in Kansas	AM isn't always about cost savings
Big Data Boondoggle	Don't automate waste
Pathfinder to nowhere	Understand the true value of AM
Cultural resistance to change – crab mentality	Not everyone will be receptive to AM
Who's in charge here?	Cross functional AM teams perform better
Engineering rigor mortis	It costs $ of waste to over analyze
Suckers for sunk costs	Develop a strategy before buying equipment
Relentless drive of organizational dysfunction	Org. structure affects AM performance
Knives out	Think carefully about AM work teams
Innovate NOW!	Innovation cannot be prescribed

References

1. Miller, C. D. (January, 2015). A Phenomenological Analysis of the Crabs in the Barrel Syndrome. *Academy of Management Proceedings*. Academy of Management, doi:10.5465/AMBPP.2015.13710abstract.
2. Shanker, A. (June 19, 1994). Where We Stand: The Crab Bucket Syndrome. *The New York Times*.
3. Tuckman, B. W. (1965). Developmental sequence in small groups. *Psychological Bulletin, 63*(6), 384–399.
4. Leonard-Barton, D. (1995). Wellsprings of Knowledge: Building and Sustaining the Sources of Innovation. University of Illinois at Urbana-Champaign's Academy for Entrepreneurial Leadership Historical Research Reference in Entrepreneurship. Available at SSRN: https://ssrn.com/abstract=1496178.

chapter five

Impact of disruptive technology

The general theme of disruption has been a major business idea since Clayton Christensen, Joseph Bower, and team coined the term "disruptive innovation" in the 1995 article *Disruptive Technologies: Catching the Wave* and later elaborated on their ideas in the 1997 book *The Innovator's Dilemma: When New Technologies Cause Great Firms to Fail*. Similar to this book, the target audience was executive management and the shift in emphasis was from an intrinsic focus on the technology itself to the business model enabled the new technology. There are two key aspects that make an innovation disruptive according to Christensen, although today the term is broadly applied outside its original definition. First, is that the innovation consists of a different combination of performance attributes, often perceived as lower quality, that initially are not highly valued by the main customer base. Second, once anchored in a niche market not profitable for the incumbent to defend, the innovators subsequently rapidly improve the attributes and eventually assault the established market [1].

There are many excellent case studies of successful disruptive innovation from the last few decades, but perhaps the best known is Blockbuster versus Netflix, which has the added benefit that many readers lived this disruption. From its humble Dallas beginnings in 1985, Blockbuster defined an industry and boasted a market capitalization of $6B with over 10,000 stores at its peak, each renting many thousands of movies and video games [2]. Netflix started in 1997 as a DVD-by-mail subscription service that was cheaper than the convenient neighborhood stores and eliminated the much-hated late fees. Blockbuster eventually responded and offered its own mail service in 2004, but it never worked. A few years later, the big shift happened with the emphasis to reliable streaming content at a flat fee and Netflix now represents the single largest source of North American Internet traffic ("Real Time Entertainment") with more than one-third of overall demand [3]. Today, Netflix has revenues nearing $12B and continues to expand its digital distribution model with original programming. Blockbuster declared bankruptcy in 2010, with DISH network hoping to realize a return on its $300 million purchase from intangible assets and digital services as it continues to close the remaining, once ubiquitous neighborhood stores [2].

Of course, not all disruption attempts are successful. This reminds us that there are more factors in play than pure technical achievement, and that the voice of the customer must provide a context for engineering innovation. We'll go back into automotive history for one powerful example of this. The Chrysler Airflow was introduced in 1934 and had a radically different and advanced design to reduce air resistance and improve handling through superior weight distribution. Reportedly, one of the more humorous findings during the engineers' research and development program was that the conventional two-box automobile designs of the day were more aerodynamic when driven backwards. Despite a theoretically much better performing vehicle, they struggled to manufacture the more complex design. Chrysler's attempt to differentiate their cars was not accepted by a public that found it too radical, and it was a financial disaster that nearly bankrupted Chrysler/DeSoto [4], barely clearing 10,000 units sold in its first year.

However, that did not stop its competitors who recognized the design advantages from copying the Airflow's advanced features. Colton's most famous quote "Imitation is the sincerest [form] of flattery" [5] applies well here and plants the seeds of a major industry shift in the 1970s. Ford, General Motors, and Volvo adopted the form and architecture successfully, and it was a strong influence on a new automotive venture called Toyoda (re-branded as Toyota in 1937). Toyoda introduced their Standard Sedan AA in 1936, and the similarities of the overall form are obvious as illustrated in Figure 5.1.

Some readers may have experienced the oil crisis of the 1970s, but most will have at least passing familiarity with its importance. U.S. design studios were slow to downsize the enormous and flamboyant (chrome and tailfins) designs that started in the 1950s [4] with models such as the Chevrolet Bel Air and Cadillac Eldorado. The fifth-generation, full-sized luxury Eldorado sedan increased in size in 1971 to over 18 ft long and nearly 5,000 lbs and was not significantly downsized and made somewhat more efficient until 1979 with the sixth generation, when a diesel engine was also offered [6]. After the oil crisis, smaller Japanese imports featuring front wheel drive and unibody construction such as

Figure 5.1 Comparison of 1934 Chrysler Airflow with 1936 Toyota Standard Sedan AA [7, 8].

the Toyota Corolla became the market (and eventually global) leader. U.S. manufacturer market share dropped from 95% in the 1970s to less than 50% in 2018 [4, 9].

Additive manufacturing as disruptor

Now that "disruptive technology" has entered the common vocabulary at companies, leaders seeking to avoid the fate of Blockbuster have a renewed focus on being disruptors rather than the disrupted. By mid-2018, a handful of Blockbuster stores were still operating (the second and third to last recently shut down [10]). This naturally leads one to speculate that similar zombie companies (or at least product divisions) must exist in many other industries, such as the increasingly commoditized commercial aviation industry, with executives everywhere racing to find the vaccine. Additive manufacturing is one of many technology megatrends – along with Cloud Computing, Internet of Things, and Advanced Automation – contributing to the Fourth Industrial Revolution or more frequently described using the buzzword "Industry 4.0 [11]." Smart leaders everywhere are trying to figure out what additive manufacturing (AM) means to their companies and how to best leverage these technologies and reverse the commoditization trend.

Arguably, the basic principles underlying AM have been around since at least the late 1800s, although the individual key enabling technologies did not reach maturity until the late 1980s. In 1981, Hideo Kodama of Nagoya Municipal Industrial Research Institute described a functional rapid prototyping system using photopolymers that constructed three-dimensional objects from layers corresponding to cross-sectional slices of the model [12]. By 1992, Charles Hull's company 3D Systems released the first commercial SLA machine and Carl Deckard's DTM Corporation produced the first SLS machine. It was originally marketed as a rapid prototyping technology that allowed engineers and scientists to quickly build accurate physical models.

With the introduction of metals and more capable polymers, the emphasis shifted from purely prototyping to production of tooling and test hardware. Over time, the conversation further shifted to production end uses and much work has been done to identify the markets where AM methods can economically compete with conventional mass production techniques such as casting and molding. Generally speaking, very high machine costs (we will elaborate more on this in Section III), and relatively low throughput have limited this to small quantities and highly complex components. These performance characteristics have motivated the strong adoption of AM in certain industries, especially aerospace and medical where they align well. Experienced practitioners have realized other disruptive capabilities with AM methods such as selective laser sintering, namely the capability to manufacture designs that are impossible to manufacture with other technologies. This includes design features such as lattice or cellular structures, monolithic structures rather than multiple pieces that must be assembled, thinner walls than are possible with castings, complex internal passages that cannot be machined, and even tailored mechanical properties to accommodate certain loads. Biomimicry of such natural structures is a common technique in AM due to the ability to digitally manufacture new structures of previously unknown internal complexity.

At many companies, this is revolutionizing the way that products are being designed and has been dubbed design for additive manufacturing (DfAM). Contrasted with other design for manufacturability (DFM) initiatives, the focus here is on the combined space of design function and manufacturing capability, rather than on cost prioritization. Many will equate the well-publicized support design rules for powder bed-based processes such as limiting hole sizes and overhangs to having performed DfAM. They do this without ever considering opportunities for the system within which the component operates, let alone engaging the customer to collaboratively ideate and implement improvements. The DfAM methodology strives to optimize the system and requires engineers to take a holistic view of what customer function, or even more ideally multiple functions, should be targeted to provide increased value. Often those less familiar with AM's emerging capabilities but tasked to implement the technology, will fall back on their DFM experiences rather than evaluate and plan their projects through the lens of true DfAM.

Various companies have even established terms to describe the compromise approach such as United Technologies Corporation's "Phase 1b" (Design for Producibility) design [13], General Electric's "Level 1 thinking" [14], and Honeywell's "Modified for AM" [15]. Recall from Saunders' Renishaw staircase of chapter two, and it is not hard to see why such terminology naturally arose. It is perhaps unsurprising that all these companies are established aerospace engine, subsystem, and component manufacturers that have internally struggled with the cultural change to an AM way of thinking.

Do not be fooled by use of the term DfAM, inside or outside your organization, when the implementation clearly fails to meet the criteria outlined above. We have seen many teams try to influence projects along less transformative paths due to reasons such as:

- *Take the "quick win"*: Elect to only modify the most egregiously wasteful design features carried over from the baseline process to accommodate the AM build rules. This approach gets less resistance in the organization because it is familiar and feels like fast progress is being made, getting the printed material on test and in the field.

- *Not enough information for DfAM*: We only have the customer specification which does not give us enough information to evaluate acceptability of the proposed system-level design changes. It takes too much time to schedule customer meetings engineer-to-engineer (perhaps they are in a much different time zone), and we do not want to bother them (or be bothered ourselves to come in early/late), plus our customer account manager is too busy and unresponsive.
- *Our customer/regulator does not understand this technology*: While similar to the "quick win", the argument is that a less drastic change is better because it is easier to explain and for the customer to comprehend, making it more likely to be accepted. True DfAM proposals will be characterized as "science projects" for which the customer is not ready or did not ask. This may have been somewhat valid a few years ago, but today most major industrial companies have at least been exposed to AM and possibly even have trusted in-house expert personnel to pull into the conversation. Regulators such as the Federal Aviation Administration (FAA) are also much more savvy about the technology, now that they have encountered multiple projects and seen largely common and rigorous approaches among the engine primes and airframers.

With these types of detours most likely awaiting you, we can promise that it will be difficult and frustrating for the team to execute that first project, and it will take steadfast leadership and toughness to maintain DfAM principles in the face of this kind of resistance.

Additive manufacturing is a disruptive technology, but only when effectively deployed. An ineffective deployment may become a false reassurance that all is mostly well, given that "the team really tried additive, but it just wasn't cost effective versus the currently supplied casting ... maybe in a few more years." We believe that manufacturing is at a key inflection point, and you would be doing your company a tremendous disservice if you accept such false reassurances. Compromise efforts will not yield the desired disruption or prevent you from being disrupted by the competitors that do not falter.

Despite all the excitement surrounding AM technologies and their obvious potential for disruption, there are a number of challenges that can limit and delay adoption or even preclude the technology from entering into mainstream production at a company. The following chapter describes the largest challenges from the authors' perspective. The chapter describes possible barriers such as certification requirements, the machine cost of ownership, make versus buy decisions within companies, lack of a skilled workforce, and the company's cultural adoption of change, in general. Each of these is barriers that prevent large-scale adoption of AM technology within industrial companies.

References

1. Christensen, C. (January-February 1995). Disruptive technologies: catching the wave. *Harvard Business Review*, 3.
2. Downes, L., Nunes, P. (November 7, 2013). https://hbr.org/2013/11/blockbuster-becomes-a-casualty-of-big-bang-disruption.
3. www.sandvine.com/hubfs/downloads/archive/2016-global-internet-phenomena-report-latin-america-and-north-america.pdf, accessed 3/18/18, p. 4.
4. Macey, S. (2014). *H-Point, The Fundamentals of Car Design and Packaging* (2nd Ed.), Los Angeles: Design Studio Press, pp. 22–24.
5. Colton, C. C. (1824). *Lacon, or, Many Things in a Few Words: Addressed to Those Who Think* (8th Ed.). New York: S. Marks. #217, p. 114.
6. www.automobile-catalog.com/, accessed 8/4/18.
7. Stern, R. https://commons.wikimedia.org/wiki/File:1934ChryslerAirflow.jpg, "1934ChryslerAirflow", Crop by Michael Kenworthy. https://creative-commons.org/licenses/by/2.0/legalcode, accessed 8/4/18.
8. https://commons.wikimedia.org/wiki/File:Toyoda_Standard_Sedan_AA_1936_Bertel_Schmitt.jpg, "Toyoda Standard Sedan AA 1936 Bertel Schmitt", Crop by Michael Kenworthy. https://creativecommons.org/licenses/by-sa/3.0/legalcode, accessed 8/4/18.
9. www.wsj.com/mdc/public/page/2_3022-autosales.html, accessed 8/4/18.
10. www.digitaltrends.com/movies/blockbuster-closes/, accessed 8/1/18.
11. www.engineering.com/AdvancedManufacturing/ArticleID/16521/What-Is-Industry-40-Anyway.aspx.
12. www.autodesk.com/redshift/history-of-3d-printing/, accessed 3/18/18.
13. Summary Report: Joint FAA–Air Force Workshop—Qualification/Certification of Additively Manufactured Parts, DOT/FAA/TC-17/35. www.tc.faa.gov/its/worldpac/techrpt/tc17-35.pdf, p. T-19.
14. https://3dprintingindustry.com/news/insights-ge-using-additive-manufacturing-additive-technologies-leader-106685/, accessed 3/18/18.
15. Citation for Honeywell MfAM Term, if needed.

section three

Additive manufacturing barriers

chapter six

Certification

Additive manufacturing (AM) has been introduced in the aviation industry at a time in history of heightened scrutiny. Given that fatigue causes over half of aircraft component failures [1], special caution must be paid by those implementing any new technology in this sector. From personal conversations, some have even quipped that if investment-casting technology were being recommended to manufacture components for critical applications today, then it would never have been accepted by internal gatekeepers such as Chief Engineers or external gatekeepers such as the Federal Aviation Administration (FAA). Any new manufacturing technology is by definition unproven and lacks field experience. The burden of qualification is on the engineering teams responsible for verifying the public safety of the products they introduce. Formalized tools and phase gate systems, along with the aforementioned Chief Engineering Office oversight, have been implemented to ensure that immature manufacturing systems are not green lighted for rate production.

In January 2015, the FAA launched an Additive Manufacturing National Team (AMNT), which was tasked with developing a roadmap for determining the needs for policy and guidance on the new technology as well as training for FAA personnel responsible for certifying projects employing additively manufactured components. Concerns centered around the expected rapid implementation of AM in commercial aviation and highlighted major challenges such as [2]:

- Limited understanding of acceptable ranges of variation for key manufacturing parameters
- Limited understanding of key failure mechanisms and material anomalies
- Lack of industry and material allowable databases
- Development of capable nondestructive inspection (NDI) methods
- Lack of industry specification and standards
- Lack of robust feedstock (powder) supply base
- Original equipment manufacturer-proprietary versus commodity type technology path
- Low barrier to entry for new and potentially inexperienced suppliers

Taken in sum, it is apparent that the aviation industry has undertaken a major push to define and address the various risks of AM technology. As highlighted in previous chapters, the abundance of different AM technologies across the spectrum of material types and energy systems has further complicated this task. One unique aspect of the FAA's task is that they will only deal directly with certificate holders, and it could be argued that supply chain risks have been exaggerated at times by those vocally and publically pursuing a vertical AM strategy versus companies with a horizontal strategy and developing a robust supply base. However, the requirements to implement additively manufactured products in aviation are substantial and will rely on a rigorous approach to standards development and an industry-wide expansion of process knowledge.

Certification of products for commercial aviation use requires compliance to regulatory requirements outlined in the Code of Federal Regulations (CFR), Title 14 – Aeronautics and Space, Chapter I – Federal Aviation Administration, Department of Transportation, Subchapter C – Aircraft. Contained in Part 25 are the Airworthiness standards for Transport Category Airplanes. There are four important sections when considering certification of new additively manufactured products:

- 25.603 – Materials
- 25.605 – Fabrication Methods
- 25.613 – Material Strength Properties and Design Values
- 25.619 – Special Factors

At this time, individual organizations must establish a standard approach to the development of specifications and design value testing to demonstrate compliance to these regulatory requirements.

Per FAR25.603, to meet the aerospace industry standards for safety-critical materials, suitability of these materials must "be established on the basis of experience or tests," "conform to approved specifications," and "take into account the effects of environmental conditions." Demonstrating this suitability entails a rigorous and complex materials test program that can amount to hundreds or thousands of test bars and components, depending on the criticality and novelty. This is a significant barrier in the industry, given that these costs can easily range north of US$500,000 if extensive high-temperature and dynamic testing are required for the target application.

Per FAR25.605, approved fabrication methods must "produce a consistently sound structure" and "be performed under an approved process specification" that is "substantiated by a test program." A process control document must be generated that identifies critical process parameters and associated tolerances. Today, these process specifications will generally require oversight and approval by the certificate holder. They may also define part-specific details such as inspection and approved post-processing methods and set acceptance test requirements, such as accompanying witness test specimens for strength or density, for example. Those wishing to manufacture parts for a Boeing or Airbus will need to demonstrate that their quality processes can satisfy all of these requirements.

Per FAR25.613, materials used in commercial transport applications must achieve statistically based strength properties and design values. These requirements are extensive and ultimately the design values chosen must "minimize the probability of structural failures due to material variability." For nonredundant structures, this requires a conservative design value be used that assures 99% of all produced material will exceed this material strength, with 95% statistical confidence. For redundant structures, where load can be distributed to other structures in the event of a failure, this requirement is 90% exceedance at 95% confidence. It must also take into account environmental factors such as temperature and moisture (corrosion). In the context of material variability, demonstrating that AM equipment can consistently produce the same metallurgical quality each time is a key factor in your operations. This will require extensive knowledge of machine factors affecting such quality, and documented control of each.

Per FAR25.619, there are conditions under which special factors may be necessary that can impact new material and processing methods. These factors are defined in FAR25.621-25.625, and must be applied where part of the structure has strength that is "uncertain," "likely to deteriorate in service before normal replacement," or "subject to appreciable variability." This variability may be from variation in manufacturing processes and inspection methods. To avoid the application of such special factors that have historically been used for some cast products, the manufacturer must demonstrate that their overall approach with respect to testing, process control, and inspection are sound.

Much improved industry guidance has become available over the last few years, through activities such as FAA and European Aviation Safety Agency (EASA) hosted AM conferences and industry consortiums generating various AM roadmaps. Respected organizations such as NASA have also released AM standards such as the October 2017 Marshall Space Flight Center "Standard for Additively Manufactured Spaceflight Hardware by Laser Powder Bed Fusion in Metals," which will serve as the background for future industry-wide guidance (MSFC-STD-3716).

Early assessments of laser powder bed fusion (L-PBF) equipment established that there are over 100 significant factors requiring process control – refer back to Figure 1.3 – to ensure repeatable outcomes in material properties and geometry. Various benchmarking studies have been performed by the National Institute of Standards and Technology (NIST), and many industry-leading companies have performed their own assessments of internal and external supplier capability. Generally, these studies have used a common test artifact across two or more AM systems, most often laser powder bed fusion equipment. In some cases, these studies have been extended to understand the variation in quality across multiple suppliers running the same equipment model, feedstock, ostensibly, and process [3].

What is not always appreciated and thoroughly understood by users and regulators are the details of the machine programming settings versus the processing "parameter set." Equipment manufacturers are implementing improved safeguards and user visibility to these machine settings which include critical variables such as the delay between the recoating step and the start of scanning – this is key step to industrialization of AM equipment, analogous to controlling the equipment model, equipment settings, and specific part processing parameters in a traditional component-specific weld schedule. Another example would be the interaction of the galvanometer jump speed with the chosen scanning strategy such as stripes versus chess board.

Today, most aerospace companies implementing AM are controlling all of these details in what is most often termed a fixed or frozen process control document. Assuring control of these factors in your supply chain is not to be underestimated and should be of great concern to regulators and the design authority. It is simple to verify that the right parameter set is being used, but how will the industry verify all the other key inputs to product quality such as the machine program settings?

Given a digital model, nearly any experienced AM service bureau can manufacture a product that appears identical, yet may not meet the full technical requirements of the component. For example, fatigue performance is very sensitive to the defect structure of the underlying material and small variations in the equipment behavior can significantly alter the size and distribution of these internal flaws. Some level of flaws, whether due to gas porosity of the powder feedstock or due to lack of fusion or local overheating, are inherent to the L-PBF process and the effect of this characteristic defect structure are already captured in the material database. However, a shift in defects closer to the external boundary could reduce the life of a safety critical product by a factor of ten or more in extreme cases. The average service bureau simply does not have the expertise or resources to do this level of assessment on their own and may only be familiar with these details by attending technical conferences.

With new machines and capabilities coming out frequently, teams will need to be conscious of the impact of these changes to previously certified product. Re-certifying most aerospace products on a new machine model has high switching costs and, at minimum, will require a comparative assessment of build quality between the two pieces of equipment. At this stage of AM industry maturity, it is more likely that an engineering validation test possibly including some of the original certification or qualification tests is needed to substantiate such a change. It may be a better choice to keep the product on the initially qualified platform, unless the business case is compelling to make the switch. Moving from a single laser system to a multi-laser system with a larger print bed can easily justify this extra engineering effort, if the part packing and volume are favorable.

Regulators are also increasingly aware of machine calibration issues, especially for the newer multi-laser machines. The industry is currently maturing processes to verify the equipment readiness for each production build, which generally entails a verification of delivered laser power (not simply the commanded power). This process is likely to extend into verification of scan field calibration if the equipment employs laser overlap zones prevalent in modern large format machines (e.g., SLM500, EOS M400 models) being used for aerospace production. Attention to this factor will ensure that the metallurgical quality of the material "stitched" together by two or more lasers is just as good as the material scanned by a single laser.

Aside from questions of machine calibration and set-up, it is also commonplace today to include numerous witness specimens with each build. Generally, these have only been various static test bars and at times only a single build orientation to represent the metallurgical quality of each lot of parts. More experienced users are also assessing the microstructure of the part versus the test bars which can differ significantly based on thermal history, such as for a uniform cylinder versus a complex, thin wall geometry. Robust inspection and quality assurance for critical applications should include dynamic property evaluation as well, whether fatigue or fracture toughness testing.

As the industry progresses into serial production, new approaches are needed to accelerate the certification and qualification of AM components. Today, there are very high barriers in terms of requirements and supporting process standards that must be demonstrated to ensure safety of passengers and aircraft operators. In all cases of FAA- or EASA-certified AM components, these have been part-specific in nature to date. As a next step for the industry, it is expected that part family-based standards will be developed as AM technology matures and extensive field experience is gained on these initial applications. To accomplish this will require a thorough understanding of the manufacturing process (effect of defects), post-processing, and inspections as applied across a broader range of products.

By contrast with the FAA, the FDA has a very different regulatory role. In its own words, the U.S. Food and Drug Administration (FDA) is responsible for [4]:

- "Protecting the public health by assuring that foods (except for meat from livestock, poultry and some egg products which are regulated by the U.S. Department of Agriculture) are safe, wholesome,

sanitary and properly labeled; ensuring that human and veterinary drugs, and vaccines and other biological products and *medical devices intended for human use are safe and effective*
- Protecting the public from electronic product radiation
- Assuring cosmetics and dietary supplements are safe and properly labeled
- Regulating tobacco products
- Advancing the public health by helping to speed product innovations" (emphasis added).

It is exciting to see any organizational charter, let alone for a government institution, that includes accelerating innovation for the public good. While 3D printing of food products is a growing trend, the application to medical devices such as orthopedic implants is the primary and most developed use case at this time. In the future, direct printing of cells and tissues will likely drive a robust public debate as these pioneering techniques continue to be adopted.

The FDA issued draft guidance in May 2016 for "Technical Considerations for Additive Manufactured Devices," which provides a broad framework for evaluating AM and supplemental guidance for devices. It is divided into five major categories of design and manufacturing considerations and motivations for control, all of which are generally applicable to engineered devices, but highlighting the AM-specific concerns:

1. *Device description*: Design intent and implementation of patient-specific characteristics with proposed AM method
2. *Software workflow*: Understand how changes impact accuracy and repeatability of the final device
3. *Material controls*: Understand how feedstock influences the physical and chemical properties of the output material
4. *Post processing*: Post-processing is an important factor to most AM processes and can significantly affect component performance
5. *Process validation and acceptance*: Processing monitoring methods, triggers for re-validation of the proposed process, and success criteria

Of special interest to AM medical devices are cleaning and biocompatibility. ISO 10993 and other guidance define the demonstration of biocompatibility requirements for any AM process. Certainly, for powder bed-based processes, cleaning of worst-case geometries such as blind holes or complex features intended to promote attachment of bone or tissues will require robust and verified methods for removing trapped or poorly adhered particles.

References

1. Findlay, S. J., Harrison, N. D. (November, 2002). Why aircraft fail, *Materials Today*, 5(11), 18–25.
2. On-Demand Mobility Workshop "FAA Perspectives on Additive Manufacturing", March 9, 2016, Kabbara and Gorelik.
3. Shah, P., Racasan, R., Bills, P. (2016). Comparison of Different Additive Manufacturing Methods Using Computed Tomography. *Case Studies in Nondestructive Testing and Evaluation*, 6, 69–78.
4. www.fda.gov/AboutFDA/Transparency/Basics/ucm 194877.htm.

chapter seven

Machine cost of ownership

Additive manufacturing (AM) equipment represents a major capital investment for any company, especially when considering the operational costs and consumables. While these machines are improving by leaps and bounds with each new model release, the productivity is still quite low compared with traditional metal-manufacturing techniques. In the following example, we will compare laser powder bed fusion (L-PBF) AM against its primary competitor in many industries, namely investment casting. This exercise is intended to highlight the critical importance of the right AM strategy. While many industrialization levers are being pulled at this stage, AM equipment still has relatively low build rates measured in only tens or hundreds of milliliters per hour.

Consider that a single 16 kg vacuum induction melting investment casting (VIM-IC) furnace costs about US\$3 million, and let us further assume annual maintenance costs of roughly 10% or US\$300,000. Such equipment would generally have a useful life of 15 years or more. Operating around the clock for a fully matured, high-rate production process, you can expect to pour every 30 min yielding 48 batches of parts per day; taking into account wasted material for gating, we will assume 24 arbitrary, 0.5 kg equiaxed Inconel® 718 alloy (IN718) parts per mold. Per year, this single piece of equipment could ideally produce 400,000 such parts, depending on the process yield, number of shifts, and uptime. While three full shifts may not be typical operating procedure at most foundries, it is being included given that the competing AM technology is commonly touted as a 24/7 process.

Let us further consider the most attractive economic scenario for additive parts which are such small (especially in the vertical Z build direction), complex components that can be densely packed on a build plate. We will assume that 48 of the same 0.5 kg IN718 components can be packed onto a large EOS M400-4 build plate of 400 mm by 400 mm (15.7 in.^{2}), along with any requisite quality specimens for mechanical property testing, powder retention, etc. As of 2018, this unit is one of the most productive AM machines on the market and capable of idealized build rates of around 100 cm^3/h (about 6 in.^{3}) due to its use of four lasers for this material. For comparison, SLM Solutions advertises the most productive SLM500 quad laser system currently on the market as capable of volumetric build rates of 171 cm^3/h (about 10 in.^{3}). Note that these are both optimistic build rate figures for print-worthy geometries and highly dependent on numerous factors which will be tabled for the moment.

If each build requires 4 days to print, then we can project that about 3,300 parts per year may be made on this single machine after taking into account yield and expected machine utilization rates. Today, the EOS M400-4 costs about US$1.5 million each with the recommended machine options. We will conservatively assume similar maintenance and extended service/warranty costs of about 10% per year as shown in Figure 7.1.

Comparing the overall productivity of each machine will inform our strategic approach to L-PBF technology. On a volume basis, we will need at least 120 EOS M400-4 printers to match the throughput of the 16 kg VIM-IC. On a cost basis, the VIM-IC is at least 70 times as cost effective. In 2018, the technology inherently is not cost competitive with a fully leaned out investment-casting process to make the same "raw" component, prior to necessary finishing operations (which AM may be able to reduce by getting closer to the desired net shape). It should be emphasized that we have ignored factory floor space, utilities consumption and supply, required personnel, and other important costs of a real factory for simplicity. Further, while the useful life of L-PBF machines is not well understood for early generation equipment and continues to mature, we believe that 7–10 years is a reasonable estimate with upgrades, so these machines could need to be replaced long before the VIM-IC requires a significant overhaul. More productive and/or lower cost equipment will continue to rapidly obsolete previous investments analogous to what happens with personal computers due to a combination of Moore's law and lower manufacturing cost with increasing machine sales volume.

The previous exercise did not address directly the recurring unit cost of the components. However, it should be obvious that the "raw" additive components will cost more when burdened with the associated capital equipment and overhead. So, why should you consider L-PBF at all? This is where Design for AM methods (DfAMs) can change the conversation from a focus on cost to one about total customer value. DfAM and design integration, properly executed, delivers greater performance, weight, and cost (often, all three simultaneously) and can meet other challenging customer requirements that may be impossible for other technologies to

Factor	VIM-IC Furnace	EOS M400 L-PBF
Capital Cost	US$3M	US$1.5M
Annual Maintenance	US$300k	US$150k
Estimated Life Span	15 years	10 years
Yield	1152 parts/day	48 parts/day
Annual Yield	400k parts	3,300 parts

Figure 7.1 Technology comparison, investment casting versus L-PBF.

satisfy. In other words, if you want AM to win a cost–benefit scenario over a traditional technology, design the part to be so complex and functional that it cannot be built any other way!

Such a design may also be delivered in a shorter overall development cycle by iterating from minimum viable products to the full-featured final product all without hard tooling costs. These tooling costs and associated lead times can quickly erase a set of business case assumptions made many months before engineering execution. AM offers a set of flexible tools and makes your organization more customer responsive through far greater agility and risk tolerance. It is a powerful lever to reenergize tired product lines and resist commoditization pressures that can start a talent death spiral in your organization.

The resulting design output should not be able to be investment cast at all due to thinner walls, complex internal lattice structures, and/ or external isogrid stiffeners. The cost of a cast equivalent may be made irrelevant by delivering weight reductions for which your customers can define valuations in terms of cost per unit weight; in some industries, this can exceed US$20,000 per pound as shown in Figure 7.2.

Combined with superior mechanical properties to cast material, the finishing costs to convert that "raw" additive part into the customer deliverable product may also be reduced. The resulting design benefits are doubly powerful when coupled with the certainty that the established machine original equipment manufacturers are already facing increasing competition as their key patents expire. The industry leaders must continue to innovate or perish and are investing very high percentages of profits back into research and development. Therefore, we can be confident that if there is a business case for such an L-PBF design today, that it will only become more attractive over time.

These authors have no doubt that for certain product families such as heat exchangers, combustor cases, and complex fluid manifolds, AM will soon be recognized as the preferred manufacturing technology. While the example focused on investment-casting technology as a comparative baseline, it also trades favorably with complex fabrications such as conventional tube-and-shell and stacked plate-and-frame heat exchangers. For many applications, it is already challenging brazed

Industry	Estimated Relative Value of 1 kg Mass Reduction
Automotive - Motorsports	$100,000
Aviation - Space	$10,000
Aviation - Commercial	$1,000
Automotive - General	$100

Figure 7.2 Value of mass reduction by industry.

plate-and-fin design – the industry gold standard for performance and weight with its 0.003–0.005 in. (approximately 0.1 mm) – ubiquitously employed within the aerospace industry.

In the case of tube-and-shell, these heat exchanger types consist of many hundreds of formed tubes mechanically or metallurgically (brazing or welding) bonded to various precision machined components such as internal baffles and the outer shell and end caps. More elaborate designs to improve heat exchanger performance or provide additional system functionality will incorporate flow turbulators on the inside and/or outside of the tubes, integral bypass valves, drains, pressure and temperature sensor ports, for example. The ability to integrate these features with an additive design approach eliminates potentially hundreds of joints and failure opportunities during both original manufacturing and in service. Even if the exact material properties and thermo-mechanical properties are not matched, the additive design provides greater value through a resilient monolithic construction that lowers the total product lifecycle cost. In many cases, the ability to pursue an alternative packaging/form factor and eliminate wasted mass due to conventional manufacturing constraints will also deliver a higher performing unit.

Equipment manufacturers such as China's Farsoon Technologies are already offering competitive quality equipment at a significant discount approaching 50% for some basic models, usually single laser systems with relatively low productivity compared to the more industrial models from the above example. We expect this trend to continue inexorably with the higher complexity and productivity multi-laser systems over the coming years. This will significantly lower the barrier to entry for more companies and expand adoption in industries beyond the core of aerospace and medical.

Environmental health and safety considerations

This section is non-exhaustive and intended to be illustrative of a significant factor that, if unmanaged, could hamper the implementation of AM technologies within your organization. The authors make no claim to deep expertise in environmental health and safety (EH&S) and encourage readers to seek their in-house experts or outside consultants specialized in this area, especially since local and national regulations may vary significantly and can be difficult to navigate.

EH&S considerations are an oft-overlooked barrier to AM adoption, especially for large companies with mature systems. Most AM experts even consider this far out of scope, preferring to focus on the more glamorous aspects of the new technologies. However, it is this author's opinion that the most thorough and effective AM experts will be prepared to handle the numerous inquiries coming from EH&S specialists. It is imperative that they should proactively equip executive leadership with the key answers to common questions such as personnel safety regarding inhalation hazards, factory cleanliness, fire safety, and hazardous waste. Executive manufacturing leaders should demand detailed control plans in each of these areas when evaluating vertical integration of the various AM technologies.

This topic is no less important when identifying supply chain partners that may not have established the same systemic controls and process rigor initially. An industry "black-eye" due to poor factory practices at a supplier can blow back on your own brand and damage the AM industry as a whole. Executive sourcing leaders must equip their supplier quality teams with the right set of questions from AM experts. Simply verifying that your supplier possesses an industry quality management certification such as ISO 9001 is a great step but is not adequate due diligence, although it can certainly provide an effective process-based approach and framework for managing the EH&S aspects of novel AM processes. The authors recommend that AM experts well versed in the specific processing technology area being sourced should participate in the initial supplier audits. We further recommend identifying supply chain partners with effective environmental management systems – ISO 14001 and its supporting standards are a well-known example – that are continuously measuring and improving environmental impact at their respective factories.

Despite the sleek machines and immaculate appearance of equipment at general industry or AM-specific trade shows, AM has much in common with other industrial metal-manufacturing processes. The AM industry may even be marketed as "clean" manufacturing technology that is inherently more environmentally friendly, given that there is likely to be less processing waste or "scrap" generated for an additive process versus a subtractive one. The metal powder-based processes are perhaps most notoriously prone to generating hazardous wastes, but similar scrutiny is warranted for handling of mixtures of uncured resins from liquid polymer-based processes such as vat polymerization.

Of particular concern is the handling of explosive and/or flammable materials in your factory. Conventional technologies such as grinding or cutting will generate combustible metal fines as well and there are well-understood criteria for such a dust explosion [1]:

1. The dust has to be combustible.
2. It must be suspended in air.
3. It must be fine enough to propagate flame.
4. The concentration of the suspended dust must be within the explosible range.

5. An ignition source contacting the dust suspension must have enough energy to initiate flame propagation, that is, combustion of the particles.
6. Enough oxygen or other oxidizer must be available to support and sustain combustion of the dust suspension.

Due to these risks, proactive measures to mitigate the spread of such powders within a shared factory space should be established such as daily vacuuming using an explosion proof design model, since conventional vacuums can greatly contribute to the simultaneous presence of the six conditions above. Alternatively, a dedicated printing space or standalone facility may be set up that will help greatly with these risk factors and ensure that only dedicated fire safety systems appropriate to the hazard are deployed (fire extinguisher class, inert gas suppression systems). In any case, a detailed plan must be in place for each aspect of managing these materials from receipt to storage to loading/unloading within equipment to cleaning of finished components and eventually recycling of powder and other scrap.

When printing heavier metals such as steels, cobalt-base, and nickel-base alloys the combustibility concerns are muted; however, exemplary shop practices are still required to minimize the inhalation hazards present and maintain employee occupational exposure in accordance with applicable limits [2]. Of course, suitable personal protective equipment must be supplied to the skilled workers any time these hazards may be present. You should not only review and understand your local regulations but also those of your customers. Some may elect or be required to manage their global supply chain risks in accordance with their own regulations such as the European Regulation on Registration, Evaluation, Authorisation and Restriction of Chemicals (REACH) or those of the California Environmental Protection Agency.

References

1. Guidelines for Handling Aluminum Fines Generated during Various Aluminum Fabricating Operations, The Aluminum Association, www.aluminum.org/sites/default/files/Safe%20Handling%20of%20 Aluminum%20Fine%20Particles.pdf, accessed 7/3/18.
2. "Proposed Cobalt Reclassification Raises Concerns for Metal Powder Users." *Metal AM.* October 2017. www.metal-am.com/proposed-cobalt-reclassification-raises-concerns-metal-powder-users/, accessed 7/31/18.

chapter eight

Make vs. buy

Any organization developing its additive manufacturing (AM) strategy eventually must define their approach for full-rate production. As covered in the previous section, the capital cost of machines has been a major barrier to AM adoption. Aside from a handful of forward-thinking companies, only the largest technology companies and a few universities made significant investments before 2015. A relatively small number of AM service bureaus have grown up globally since the early 1990s, many of which started with rapid prototyping services in various polymer materials, especially the stereolithography visualization tools or fit check hardware now commonly found in engineers' cubicles everywhere. Others focused on repair technology, especially laser cladding or deposition to rebuild worn surfaces. In recent years, a few of these service bureaus have been positioning themselves for production and now advertise themselves as manufacturing partners with significant factory footprints. It should go without saying that production requires production behaviors, and current generation equipment still has a great deal more rapid prototyping in its genes than rate production.

Accordingly, most of these potential suppliers had to develop in-house expertise in metals manufacturing of new products, most commonly the powder bed fusion technologies of laser powder bed fusion (L-PBF) and electron-beam melting (EBM), but also binder jetting and directed energy deposition (DED). Some of these companies already had conventional manufacturing facilities and an existing customer base, typically high-end machining or casting, but others were new market entrants, usually specializing in a single AM technology or even equipment manufacturer (i.e., a supplier might have 6–8 Arcam EBM machines, but only 1 or no L-PBF equipment vs. another supplier with only EOS L-PBF equipment).

Given the hype around AM as one of the drivers of the next industrial revolution, many of these service bureaus have been acquired by companies looking to secure capacity and expertise in the still emerging technology. Perhaps the most widely known example of this is Morris Technologies, Inc. (MTI) and its acquisition by General Electric Aviation in November 2012. Prior to this, MTI represented the largest metal L-PBF capacity with roughly 20 machines, at a time when the global installed base of metal printing equipment was only a few hundred [1]. With CEO Greg Morris' forward-thinking investment came a focus on maturing the technology for production and many would cite MTI's modification of the equipment to enhance the mechanical properties and development of improved processing parameter sets as a key step toward industrialization of AM. Notably, they also had team members skilled in co-developing designs with customers that could best leverage the technology's unique capabilities and minimize the impact of its oft underestimated post-processing downsides.

As a business leader evaluating an investment in this technology, you are bombarded by examples of how your competitors are applying this technology from news articles and prototypes on display at tradeshows. How will you approach key decisions on how you will implement AM in your business given the myriad technologies available, variety in equipment manufacturers, and competing voices within your company? This chapter will share the best practices and lessons learned on vertical versus horizontal sourcing strategies ("Make vs. Buy") and the relative merits of each as a function of your company's AM maturity and needs. (https://wohlersassociates.com/blog/2012/07/morris-technologies, accessed 3/20/18.)

Which technology?

In Section I, we outlined the seven AM process families as recognized by the ASTM International Technical Committee F42 on Additive Manufacturing Technologies. Furthermore, it is important to recognize that not all these technologies are at the same maturity level with respect to production. The U.S. National Aeronautics and Space Administration (NASA) developed a common framework for assessing the maturity of a technology known as Technology Readiness Levels (TRL). Since its formal introduction in 1989, this basic methodology has been broadly adopted with modifications by many organizations such as the European Space Agency and U.S. Department of Defense, and accordingly, it is also popular at most industrial and technology companies. Refer to Figure 8.1 for a brief highlight of the key characteristics of each of these stages [2].

Generally, the most significant steps in technology development are achieving TRL3 and TRL6. At TRL3, the basic "proof-of-concept" has been demonstrated and the benefits and promise represented by the technology can be realistically assessed.

This is a key decision gate, as the expenses involved in further developing the detailed business case and identifying a specific application and overall implementation strategy grow quickly in the following engineering project phases. As the project progresses to TRL5 and manufacturing of validation hardware and bench testing, significant funding and personnel resources are committed with costs typically ranging into millions of US$ for new technologies. To achieve TRL6 at a turbine engine business

TRL #	Key Characteristic
TRL 1	Investigators are demonstrating **basic principles** of a new research and technology area.
TRL 2	Investigators have established a **concept application** for the new technology.
TRL 3	Investigators achieve TRL3 upon completion of a simple **proof-of-concept** that demonstrates the new characteristic or function.
TRL 4	Investigators validate the concept application as a **minimally viable product** in a controlled lab-scale experiment.
TRL 5	Investigators demonstrate the key risks and functions of the **detailed design** in a high-fidelity bench test.
TRL 6	Investigators update the detailed design for a **system level test** in the actual operating environment.

Figure 8.1 Technology readiness level key characteristics.

for a new product, the required ground engine test alone will exceed US$1 million with 150 h of endurance testing as a minimum expectation. Considering these risks, it is easy to see why industry adopted a formal framework with specific phase exit criteria for evaluating new technologies originally developed for purposes of space exploration.

As of mid-2018, the various AM technologies have been broadly adopted for production end use by the aerospace and medical industries, with some limited automotive usage, especially in motorsports. The ASTM F42 process categories are extremely broad and must be broken down for proper assessment of industrial application maturity and potential. From the perspective of aerospace, nearly all AM modalities have already achieved at least TRL6 and are being used in low-rate production; Boeing completed the F-15 pylon rib using a directed energy process for a titanium alloy in 2004. In light of the barriers to certification, it is clear that industries with lower regulatory requirements can rapidly mature TRL of a technology for their desired use. It should be noted that with any summary of equipment manufacturers versus AM technologies, such as Figure 1.1, is not intended to be an exhaustive list. There are multiple new market entrants announced each year, increasing competition and lowering costs.

Today, L-PBF and wire-fed DED are the dominant production AM metal technologies that can produce mechanical properties consistent with aerospace industry requirements. As mentioned in Section I, wire-fed DED is most commonly used by aircraft manufacturers (airframers) for high value titanium and nickel alloy structures to reduce the buy-to-fly ratio and cost of these components. L-PBF is more broadly applied due to its strong combination of feature resolution and good mechanical properties, allowing it to be used for some engine structures and mechanical systems such as pneumatic valves, heat exchangers, and actuators. EBM also has some well-publicized components in production, especially GE Aviation's low-pressure turbine blades made from a titanium–aluminide alloy that cannot be printed by L-PBF.

To implement similar applications in your own business you will\ need either to purchase equipment of your own and develop the associated infrastructure and personnel or to work with suppliers that have these capabilities. Many aerospace companies have opted to in-source these capabilities due to the treatment of the technology as "new and novel" by the Federal Aviation Administration (FAA), other regulators, and customers uncertain about the structural integrity of the resulting metal parts relative to conventionally manufactured parts.

Despite this, the industry has matured substantially in its understanding of the interaction of process of parameters and resulting material properties. Further, the L-PBF equipment of 2018 are more capable than the earlier offerings which were not suited for serial production without significant modification and process monitoring. Significant capacity has been established in the supply base and is steadily growing to accommodate the shift to production from prototyping, coupled with greater awareness of the associated increase in quality requirements.

This most commonly proceeds in a laboratory model, making prototypes and tooling initially. Many teams have elected to then procure metal prototypes to illustrate the potential of an AM technique to make a desired new concept model that cannot be traditionally manufactured. Next, strategic partnerships are formed after evaluating the service bureaus' capabilities to meet costs, quality, and other typical factors. The initial goal of such a partnership is to determine how to go about industrializing the technology as part of the TRL maturation process in light of the regulatory burden. At this point, a key inflection point is reached and decisions must be made to either expand said partnership or acquire equipment and personnel, whether organically or by acquiring one or more service bureaus. The decision to operate an internal AM facility geared toward full-rate production should not be taken lightly and is often motivated primarily by the need to assert complete process control over the printing and post-processing of an item in what is most often termed "fixed process" or "frozen process" control; rather than controlling the output quality of the process, all relevant inputs are also defined, documented, and controlled. As indicated, there are a few options for how these details could evolve – or more likely, are already evolving – at your company, but this is generally a common low risk approach.

For less regulated industries and/or start-up companies, taking the major step to invest in AM equipment may diverge significantly from a risk-averse aerospace approach. The authors have asserted that, unless an undertaking is developing new machine technology, unique manufacturing strategies, novel materials developed for an AM process, or some combination thereof, the primary value of AM resides in the product design being improved through rigorous DfAM. It may make little sense to internalize the challenges of running an AM manufacturing facility as this would only dilute focus from your core value creation. If there is a strategic need to bring these capabilities in-house, then the selected equipment should be appropriate to the problem being solved.

Investing in AM equipment is much bigger than just buying a few new machines. Some other important factors to incorporate into your strategic decision-making include:

- *Workforce*: Due to its criticality, we address this in the "Skilled Workforce, Or Lack Thereof" section at length. Ensure adequate staffing to turn over the machines promptly and work closely with your equipment provider's field service engineers to diagnose problems. This is not terribly difficult in the early days, but will rapidly get out of control as you scale your operations. The team will need to be able to deal with minor issues on its own to maintain high operational efficiency. Do not underestimate the relative importance of post-processing and quality inspection of parts when transitioning from prototyping to production. There is a tendency to deal with these parts of the value stream last, but they are critical to success – ensure that your team is giving them the right priority by assigning an owner. Similarly, when assessing a potential supplier, you should not neglect thoroughly evaluating these areas of their capabilities.
- *Material*: Evaluating incoming feedstock quality and storing the material, especially with fine powder, can be an expensive undertaking in some locales. We go into more detail on this in the "EH&S considerations" section of chapter seven, but depending on what material is being processed, facilities with inert gas fire suppression systems and rated containment space for reactive powders may be required. Advanced laboratory equipment to assess the composition, particle size distribution and flow behavior of the powders will not only require capital expenditures but also qualified staff and training to run the new equipment.
- *Training*: AM equipment is continuously advancing with new capabilities and systems coming online all the time. Your team will require training to use these new functions or may have to grow to realize value from a new capability such as melt pool monitoring or other integrated, in situ process quality system. As the leader in this

rapidly changing field, you will need to ensure adequate space and funding to develop your team members.

- *Facilities*: AM equipment requires utility connections for inert process gas lines, cooling water, power, and network data connections. More advanced facilities making the "Factory of the Future" a reality today will require considerable thought and planning around AM process and post-process automation. A qualified and experienced facility engineer is a key part of the AM team. Process waste will also need to be owned by someone; recycling of the fine powders, support material, and filters is not trivial.

- *Location*: The value of a distributed network of manufacturing centers in your operations versus a traditional large, centralized manufacturing plant may be worth consideration. By placing AM resources near the point of use for your customers, you will be able to respond quickly to requests and save on delivery costs. Different customer and product needs may mandate different equipment, post-processing and inspection strategies and materials being printed. Some degree of specialization can add value by not trying to have one facility serve every customer and thereby burdening steel production with the safety and regulatory constraints of aluminum, for example. On the other hand, large capital expenditures for batch processing equipment such as heat treatment furnaces may contraindicate such a strategy and drive one to consider whether it is truly necessary to vertically integrate these processes.

- *Financing*: Solid lease financing options are available and a good option for many companies, but ensure that you account for the cost of periodic system upgrades in your make versus buy analysis. For example, the gas flow systems of many machines have been significantly improved to remove the process "smoke" more efficiently to deliver better material quality. Eventually you may also realize a need for higher power lasers – from the more typically 400–500 to 700–1000 W – to improve the printing rate of your equipment by allowing an increase to the build layer thickness. Today, such new lasers are more than US$30,000 each and could be a significant expenditure if you have more than a few multi-laser systems. Filter systems are also an area seeing upgrades from disposable to permanent designs.

While the capabilities of the newer large format printers are exciting, if your company is focused on developing new AM materials that your materials specialists create, this can be performed most efficiently in a small development platform which will allow you to turn over the machines quickly with less operators and technicians and supporting infrastructure. Sub-scale, product relevant work can still be performed in such equipment and full-scale activities may be outsourced or acquired later once a base of capability is established. There are many underappreciated challenges and risks by those looking to adopt AM so strategic alignment of the true equipment capabilities and day-to-day operational realities with your goals is needed.

The current generation of machines are trending toward an industrialized product with all the associated expectations of reliability, uptime, and cost productivity, but are not yet in the same class as mature products such as CNC machining equipment. The ratio of service costs to the installed equipment base is most telling in this regard. As of December 31, 2017, SLM Solutions had 56 employees involved in after sales per its 2017 annual report with sales of 113 total machines (130 in 2016 and 102 in 2015). Sales of machines generated €73M, with €9M in after sales (includes powder sales), but nearly €20M in service-related expenses – a direct driver of a net loss for the year (https://slm-solutions.com/sites/default/files/e_slm_solutions_group_ag_annual_report_2017.pdf, accessed 9/15/18). By contrast, one publically traded machine tool manufacturer Hurco Companies, Inc. disclosed in its 2017 Annual report sales of US$209M in computerized machine tools vs. US$32M in Service Parts and Fees (https://hurcocompaniesinc.gcs-web.com/static-files/dc9968ae-4c34-400b-9cd6-d855cdfe0177, accessed 9/15/18), a far different proportion. It bears repeating that services and maintenance of any in-house AM equipment should be a primary consideration.

Another underappreciated risk is the ability to manufacture a desired shape with an appropriate AM process. While teams are undoubtedly innovating solutions to EBM equipment's sintering of powder in internal features, the removal of such partially sintered material represents both cost/waste and drives additional post-processing requirements. It also imposes fundamental design limitations that may not be aligned with your goals of say, manufacturing printed heat exchangers. This EBM example may seem trite, but more subtle considerations are familiar to practitioners working on the same printed heat exchanger challenge via L-PBF, which does not have this "caking" problem.

Thin wall structures may be manufactured using this technology as any equipment vendor at a trade show will be proudly exhibiting impressive lattices. However, what may not be immediately obvious is that such thin wall structures require entirely different laser parameters that can dramatically impact the cost productivity of L-PBF equipment. Speaking generally, these "contour" parameters are on the order of 2–3 times slower than "bulk" parameters used for thicker sections. Therefore, if you mistakenly assume that the advertised build rates in the manufacturer's brochure or website would be applicable to the product you are developing that incorporates such thin features, then your strategic goal may not ever be realizable, or at least significantly delayed due to a huge design-to-cost gap.

This gap can be addressed by your design teams in many ways, but most of the solutions are unpalatable, resulting in either degraded unit performance (i.e., make the walls thicker), higher development costs to further improve the contour parameters, or higher capital and operational costs as more equipment than expected to meet capacity or larger and more complex equipment is needed to meet the business case. Even relatively experienced AM users may not appreciate this distinction or factor it into their design decisions at an early enough stage. They are instead surprised and disappointed when it takes much longer to print any particular design. This type of experience will hinder the adoption of AM products in your company and enable a culture of resistance to necessary change.

While my example focused on the initial unit manufacturing cost, there are many other important considerations particular to heat exchangers not appropriate to this book or not shareable at this time – let it simply be said that while this is a complex AM problem, this author believes the performance results are worthwhile for early adopters.

References

1. Gorelik, G. (2017). Additive manufacturing in the context of structural integrity. *International Journal of Fatigue, 94,* 168–177.
2. Deutsch, C., Meneghini, C., Mermut, O., Lefort, M. (2010). Measuring Technology Readiness to improve Innovation Management, *Proceedings of The XXI ISPIM Conference.* Bilbao, Spain.

chapter nine

Skilled workforce, or lack thereof

To date, the current model for training for additive manufacturing (AM) technology has been left largely on the shoulders of the machine sellers. Therefore, the employees from companies who bought machines benefited in the training on how to use these machines. This is a reactive strategy to training needs and is predominately hindering the expansive deployment on how AM machines operate and limiting the amount of skilled workforce capable of running this type of equipment. I was fortunate early in my career to work for a company that made large investments in AM capital equipment, thereby giving me exclusive access and insight on equipment operation and maintenance at a technician level.

However, in order to grow the field of AM by growing the number of technicians available to run equipment, AM machine technical training must go beyond the exclusivity of industrial companies that happen to own AM equipment, but rather, deploy into vocational schools, technical colleges, technical organizations, etc. What has prohibited this strategy thus far? Original equipment cost is the largest contributor to exclusivity. AM metal machines capable of producing quality hardware currently cost anywhere from $500k to $3.25M USD, depending on the specific AM metals technology and scale of the machine. This type of sunk cost capital equipment investment has generally favored industry and neglected less funded academia . In addition, the annual maintenance cost of these expensive machines contributes to a barrier of entry for small technical schools as well. It is obvious that without an AM machine for future technician to learn off of, how can the vocational schools offer a program to generate growth in the field? So, as a result, it is often on the shoulders of industry to bear the cost of training technicians capable of running these types of machines. Student technicians are often sent to Europe for an immersive training experience with the machine manufacturers to understand and use the technology.

The above content captures a general theme of technician skillset shortfalls due to high capital equipment costs; there are exceptions to the rule, and hopefully, we are trending in a more proactive direction. In 2014, an initiative, headed by Oak Ridge National Laboratory (ORNL) [1], highlights one such proactive training model of training veterans to become AM technicians through a training program designed by ORNL in collaboration with Pellissippi State Community College.

"Funded by EERE's Advanced Manufacturing Office, Pellissippi State developed the curriculum and ORAU recruited participants from across the country. The students – 14 veterans, three active duty military personnel, two reservists and five civilians – completed three weeks of classes at Pellissippi, two weeks of hands-on training at the ORNL Manufacturing Demonstration Facility (MDF), and one week of carbon fiber composites training at the Science & Technology Park on ORNL's main campus.

ORNL engineer Dr. Lonnie Love, who led the program at MDF, believes training veterans in emerging technologies can help meet U.S. demand for a skilled manufacturing workforce." "There are literally thousands of veterans entering the civilian workforce each month," Love said. "I've been so impressed with this inaugural class, and I love to work with our veterans. Our industry partners are looking for a highly skilled workforce, and the veterans already bring to the table the discipline, teamwork, and commitment needed to develop these very specialized skills." (2014, DOE website: www.energy.gov/articles/energy-department-trains-veterans-advanced-manufacturing)

Since the start of ORNL effort, another Veteran training program has kicked off called 3D Veterans. Sponsored by America Makes, The U.S. Department of Veterans Affairs, Google and Autodesk, this association trains Veterans, free of charge, from around the United States looking for work to become AM technicians. This is a terrific example of directly addressing the technician skillset shortfall in the United States with a practical approach that creates hardworking, level-minded, experienced professionals to move AM technology forward.

Addressing the shortage of training for AM technicians aside, there is also a massive shortfall in engineers trained in AM metals technologies coming from universities. Due to the upfront capital financial commitments listed above, universities typically have also found it difficult to purchase AM metals machines to train engineers at the university level. Universities, for the most part, have done a good job at acquiring smaller and less expensive AM polymer machines that students that mechanical engineers might use in a senior design class. Of course, there are some exceptions that come to mind with the University of Louisville, University of Texas-El Paso, Arizona State University, Penn State University, and Missouri University of Science and Technology having AM metal lab capabilities. General Electric is doing its part to fix this problem. In 2017, GE set aside $8M USD for the GE Additive Education Program, which donates metal AM machines to universities around the globe [2].

Engineering departments within universities also report that there is a shortfall in metallurgical engineering as a major within universities across the United States. For example, in the fall of 2015, ASEE reports [3] that of the total amount of 26,839 tenured or tenure-track engineering faculty in the United States, only 987, or a mere 3.7% were Metallurgical &

Materials Engineering faculty. By comparison, 5,972 were Electrical/ Computer Engineering faculty and 4,808 were Mechanical Engineering faculty. Despite Mechanical, Electrical and Computer Engineers, all fulfill roles related to the AM industry; a major obstacle in the growth of AM metals is in Metallurgical and Materials Engineering field. With Metallurgical and Materials engineering lacking the popularity among young engineers, there will continue to be a shortfall in the area with respect to AM.

As discussed, once newly minted engineering graduates enter the workforce they will predominately pick up an AM skillset within industry using AM. Those new company employees will struggle with how their company is adopting the use of AM for its product lines. Depending on the company, the amount of internal corporate training available to employees will be small to perhaps even non-existent. AM consultancy companies have sprung up seemingly out of nowhere to address this need with some consultancy companies partnering to deliver training content to industry. For example, the Barnes Group Advisors partnered with Purdue University's College of Engineering to offer online AM certificates.

Some industrial companies have seen the need to address the training gap, and they have proactively created internal training programs that address how to design for the technology, how to manufacture parts with AM technology, and how to implement new approaches in product development with AM technology. Sadly, many companies have not been able to be so thoughtful in their approach. Many companies realize that they need training in the technology field and quickly if they want to stay competitive in their product designs. But perhaps they need to start with hiring an AM expert first. Herein lies an even larger challenge ... finding an AM subject matter expert to hire.

AM is a relatively new technology field, and it is very hard to find someone who has a number of years of direct experience running AM equipment, designed parts for the technology, bridged manufacturing and process challenges, installed AM equipment successfully, created and implemented training curriculum and created roadmaps that influence company strategy around AM technology. In fact, it has become so challenging to source these individuals, entirely specialized recruiting companies have started that specifically match AM experts with company needs. A large part of the AM specialized recruiting companies' skillset is the ability to filter through the vast number of pretenders and journalists in the field of AM simply recycling LinkedIn articles of other people's AM success and narrow down to the people actually involved in direct action in the technology field. Most of the AM action types are too busy working in the AM field to populate LinkedIn with AM content.

In fact, Dr. Bob Bradley, a well-known AM industry expert from the United Kingdom has had a conversation with this author on this very topic.

During the conversation he stated, "there are now too many charlatans in the field of AM, people not passionate about the technology, not engaged in moving the industry forward, just people looking to exploit the excitement of the technology and make a quick dollar." One word of advice for newly hired engineers within the AM industry is to adopt the motto I developed at my former company to motivate the AM researchers. "Make the news in AM, not report the news." Hopefully, as AM technology matures and the skills gap narrows, there will become more AM people capable of advancing the technology. America Makes has done its part in this area by acknowledging the multiple levels of training necessary. They have developed a model that identifies the learning model for the factory of the future, which is illustrated in Figure 9.1.

As discussed previously, sourcing the talent is a massive challenge. Once these people are brought in the company, you need to make the role challenging enough that this precious talent will want to stay. For example, I have seen AM experts strategically sourced for companies and left to do little AM work and often ignored within companies. After speaking to AM recruiters, I have been shocked to hear how common this problem of misuse of AM talent is within industry. Indeed, recruiters do not complain about this trend as they see it as more revenue for themselves. It is troubling that mismanagement of talented professionals is so prolific within industry that it creates growth for an entire recruitment industry to thrive.

You may be asking the logical question, why had the talent been underutilized? Well, this is often due to the dynamic politics that exist

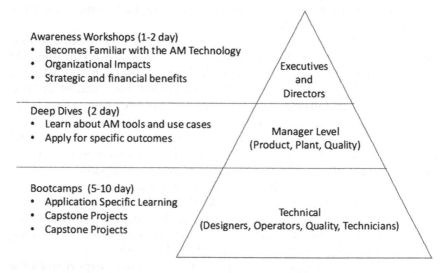

Figure 9.1 Multilevel AM training.

within industry. Political turmoil creates a number of conditions where underutilization of AM talent is rampant within companies. For instance, there may have not have been an executive level sponsorship of the need for AM before the experts were hired. Perhaps there was merely one Vice President level advocate of AM, and he/she leaves the company after a few weeks after the AM experts are hired. Since the remaining executives did not hire the AM expert, they are unaware of what to do with them and ignore them.

Or, as an alternative example, perhaps the manager hiring the AM expert underestimated the company culture change requirements necessary to properly deploy the technology within the organization and quickly loses interest in transforming AM projects thereby shuffling the AM expert into a different role. So, what can be done to avoid the above scenarios?

Make sure the entire company is truly "bought in" to the technology and that starts at the top with the highest-ranking executive. A vehicle to ensure appropriate linkage among all levels of the organization is Hoshin Kanri. We will discuss Hoshin Kanri in more detail in later chapters.

References

1. www.energy.gov/articles/energy-department-trains-veterans-advanced-manufacturing.
2. 3dprintingindustry.com/news/american-and-european-universities-benefit-from-ge-additives-1-25m-in-metal-3d-printers-135116/.
3. www.asee.org/papers-and-publications/publications/college-profiles/15En-gineeringbytheNumbersPart1.pdf.

chapter ten

Cultural adoption of change

There is no doubt that additive manufacturing (AM) inherently is a catalyst for change within organizations. At the highest level, AM change affects almost all aspects in a modern manufacturing company. AM change comes in the form of radical design practices, changes in organizational structuring based on departmental functional integration, changes in procurement strategies and inventory control, changes in quality and inspection, changes in material development, changes in repair and overhaul tactics, and changes in production line tooling. This is only a partial list of the types of changes that can occur within an organization due to AM.

Change management has been a topic of interest by organizational behaviorists for a number of years. Some of the pioneers in this field include Kurt Lewin, Rich Beckhard, and Willian Bridges. During the 1990s, businesses started to adopt "change management" within the business common language. Modern-day researchers in the area of change management include John Kotter, Daryl Conner, Spenser Johnson, and Jeanenne LaMarsh.

To give a quick synopsis of the research in the field of change management in the last half century, there are a number of factors that provide a resistance to change within organizations. These factors are listed in Figure 10.1.

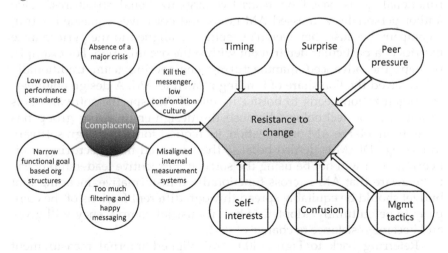

Figure 10.1 Resistance to change.

Timing, surprise, and peer pressure represent general resistance to change by employees. Not specific to AM necessarily, the squared boxes of bad management tactics, confusion, and self-interests are change specific reasons for resistance. These are often exhibited due to poor communications strategies rolling out new technology within companies.

Perhaps common across the aerospace industry, one of the most common AM functions to resist change is industrial complacency. The authors have personally witnessed this behavior on a number of occasions throughout their careers. If you look at the specific reasons why an overall sense of complacency exists within industry, there can be a collection of root causes.

Because the authors have personally witnessed these root causes throughout our careers with respect to AM, it is critical to focus more on these topics. The first of these root causes of industrial complacency would include "too much filtering among layers of management" creating a sense of happy messaging to the highest-level executive. You can almost picture in your mind an executive briefing by 25 middle- to upper-level managers in a conference room. After receiving glorious praise from the managers on the state of the company, the executive asks "Why do we need to implement AM, when we aren't experiencing any issues with our current manufacturing strategy?" Knowing that perhaps the executive has a "Kill the Messenger" type of attitude toward negative news, this creates a low confrontation culture. Despite management seeing how competitors are using AM to take weight out, costs out, and lead time out of new product introductions, often management will bury their head in the sand and keep the messaging positive.

Another root cause for industrial complacency resides in "narrow functional goals based on restrictive organizational structures." One author personally witnessed AM goals and objectives squashed within a company because they weren't specifically aligned to the performance metrics of a particular department within the organization. For example, perhaps the advanced manufacturing engineering team traditionally has received the lion share of funding related to AM. A design engineering organization needs to bolster its use of AM-specific software tools to advance the technology. Because the design-engineering group has never been a large AM organization, its request for new design software to leverage DfAM is denied because the advanced manufacturing engineering team will not be using the software. Executive leadership needs to be aware that AM is cross functionally transformative; meaning that budget should be equitably spread through different factions of the company to be culturally adopted, if not, industrial complacency will grow and become resistance to change of AM.

Referring back to Figure 10.1, "misaligned internal measurement systems" are also predominately found within the aerospace industry, which leads to more industrial complacency. As companies are acquired, merged, or divested, the cultures within these sub companies become small microcosms of culture that differs from the mother company. As such, misaligned internal measurement systems create confusion and inefficient use of resources. For example, the advanced manufacturing engineering team mentioned above may have goals to directly convert a specific number of existing conventionally manufactured parts to AM parts. This goal may be completely different than the design engineering teams goals of creating a specific number of new products solely designed for AM. These differing priorities create internal confusion, and ultimately, industrial complacency.

Think of an example when you may have a specific site within your business that has historically underperformed for many years because the product is now basically a commodity. It is not hard to imagine that "low overall performance standards" can be set by management for that site. This will lead to industrial complacency with no attention given to new and emerging technologies, such as AM, to overhaul and rescue the commodity product line. Once a product line reaches this status, it is very difficult to reenergize the team and turn it around effectively. A major contributor to this is that many of the leaders that originally developed the technology will have already left or be at a career stage where such innovation is personally undesirable, given that it is much simpler and less risky to do things the same, safe and proven way.

In our opinion, the largest root cause of industrial complacency resides in the "absence of a major crisis." A common question arises often among AM practitioners at AM conferences; how did your company implement AM into production? On many occasions, the answer has been, "we were stuck and we had no other way of manufacturing this product." If you read between the lines of this statement, basically, there was a major crisis that had befallen our company and we had to act. It is well known that with AM parts come exhaustive testing, exhaustive investments, and a lot of effort. So, an executive leader must ask his or herself, if your company has an "absence of a major crisis" what motivation does your company have to move from industrial complacency to implementing AM? This is why the next chapter is titled, "Establish a Sense of Urgency."

In addition, this brief chapter only skims the surface on cultural adoption of change. Much has been written by a vast number of organizational behaviorists over the past 60 years or so. To gain a broader picture on this topic, we encourage the reader to dig through some terrific papers on workplace change management and new technology insertion. These research data seem to age quite well with little noticeable differences from the 1960s to present day on the topic.

section four

Applying change management best practices to additive manufacturing

chapter eleven

Establish a sense of urgency

In order for any new initiative to gain momentum, there must be a sense of urgency or burning platform. This is the same for all corporate level change efforts regardless if it is additive manufacturing (AM) or Lean or Six Sigma. There has to be a recognition that in order to move the company forward, things must change. If status quo is acceptable, then it will be difficult to gain the necessary buy-in for AM adoption. This chapter discusses the importance of and methods for establishing a sense of urgency around AM.

Leading change

John Kotter [1] developed an eight-step model for leading change, which is provided in Figure 11.1. The next several chapters will discuss how to lead an organization through the change of adopting AM.

Leading AM change throughout a company can be challenging. Management should ask several questions when deciding which AM change efforts are worthwhile:

- What is the evidence that the approach really can produce positive results?
- Is the approach relevant to your company's strategies and priorities?
- Can you assess the costs and potential benefits?
- Does it really help people add value through their work?
- Does it help the company focus better on customers and what the customers value?
- Can you go through the decision-making process, understand what you are facing, and feel that you are taking the right approach?

A PICK (Possible, Implement, Challenge, and Kill) chart is an easy way for management to categorize AM change ideas. A PICK chart is a visual tool for organizing and prioritizing ideas. Often after brainstorming session, a PICK chart is used to determine which ideas can be implemented easily with a high payoff. Figure 11.2 provides a PICK chart.

From a management perspective, a PICK chart enables management to compare ideas based on their impact and ease or cost of implementation. Ideas that are easy to do and have a high payoff should be implemented.

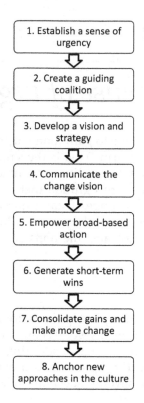

Figure 11.1 Model for leading change [1].

	Low Payoff	High Payoff
Easy	Possible	Implement
Hard	Kill	Challenge

Figure 11.2 PICK chart.

Ideas that management should consider as possible projects are those that are easy to do but have a low payoff. Ideas that have a high payoff but are hard to do will be a challenge. Management should consider these carefully to determine if they have the resources to accomplish these. Finally, ideas that are hard to do and have a low payoff should not be pursued.

Establishing a burning platform

The first step to leading change is to establish a sense of urgency, or burning platform, within the company itself. Leaders must ask as agents of change in order for change to occur. Creating a sense of urgency is often accomplished by senior leadership through actions that capture attention from organizational stakeholders and communicating the need to quickly respond to a condition. By explaining the need, leaders alert the organization of the need to change and begin preparing all levels of the organization for the change process. The key is communicating the urgency because this leads to cooperation of the affected stakeholders and meaningful change. It is the first and most important step in leading change.

It is not sufficient for leaders to simply request action. Leadership must provide a vision for the value of the future state in order for organizational stakeholders to buy in. Further, leaders must make it clear that maintaining the status quo is not an option and explain why. There must be a compelling narrative that leaders provide to show why the organization cannot remain in its current state. Leaders cannot shy away from open and honest discussions about the current market, competition, and customer environment as well as relevant financials. Data supports the urgency. A lack of transparency, however, can lead to reduced buy-in.

In many cases, your company may be responding to an already existing crisis that necessitates the need for AM adoption. For example, perhaps a competitor is advertising a competing product's use of AM technology. Or, perhaps your company has lost its sole source supplier of a particular component and the production line will shut down if a part is not available quickly, which was discussed in previous chapters, AM may serve a role in this scenario.

Conversely, your company may only be dabbling in AM technology. Perhaps your company is merely investigating where the technology might fit within your product lines, or perhaps how it may save lead time and cost for your company without a clear goal in mind. In this case, you may need to advocate for a sense of urgency to use AM within your company. Many change agent theorists believe that in order for a new technology to gain a foothold within companies, there needs to be a sense of urgency to deploy it. If no urgency exists, then a lower risk base will naturally hedge toward known technologies with established track records of manufacturing and design performance.

For example, you overhear a conversation between a project manager named Sara and a structural engineer named Johal. Sara claims that management has been lobbying for more AM integration in design; however, Johal seems convinced that AM is a riskier approach because Johal does not have three decades of mechanical property knowledge about how AM will behave from a structural performance standpoint. Johal points out that in order to develop the confidence in the properties, a large battery of destructive testing, that will take months to complete and over $1 million dollars of testing, will need to be performed to establish a basis for Johal to cover his ass. Johal exemplifies Elon Musk's quote of "There's a tremendous bias against taking risks. Everyone is trying to optimize their ass-covering." Sara takes Johal's message of restraint back to executive management, who not knowing better, take Johal at his word and kills any AM initiative due to being cost prohibitive.

Now, alternatively consider if the product that Johal is responsible for analyzing is being discontinued due to lack of sales and now Johal will find himself without a job in the near future because a competitor released a higher performing, lighter weight, competing product. Do you think that Johal could find an alternative way to certify an AM offering in those conditions? Perhaps quickly building and destructively testing an AM component or components to empirically feed into structural analysis models? Johal would probably find a way given Johal's job is now on the line. This scenario is what is meant by establishing a sense of urgency.

So, how does one go about communicating a sense of urgency? Luckily, there are several steps that leaders can use to create a sense of urgency and increase buy-in. First, leaders should show their commitment to the change by providing adequate resources, training, and time. In order for AM to be successfully adopted, employees must be trained and given the appropriate equipment and tools. By outlining a strategy for rolling out an AM training program, the leaders are showing their support and commitment. It also indicates that this will not be a passing fad, rather it will be a new business approach. Second, leaders must have open lines of communication. This means sharing the good news and the bad news.

There are several specific approaches to enlisting AM cooperation, which will enable a smoother implementation. These approaches include education and awareness, negotiation and reward, manipulation and co-optation, explicit and implicit coercion, facilitation and support, and participation and involvement as shown in Figure 11.3.

In order for an AM implementation to be successful, management must offer education and increase awareness of AM technology. First, training programs on AM must be rolled out that are targeted for employees at all levels of the organization. For example, while finance is not directly involved in production, they will need to understand the impact of AM on supplies, materials, and equipment. Organizations should also consider multifunctional technical and nontechnical-based certifications to encourage interest in increasing AM knowledge.

Management must also consider employee participation and involvement. Companywide initiatives to utilize AM through linking implementation to department metrics and goals will show management's commitment to AM. For example, organizations can set goals which

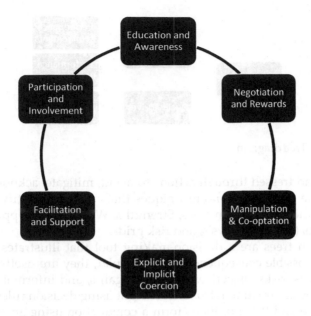

Figure 11.3 Approaches to enlisting AM cooperation.

require a certain monetary target of value add using AM per new product development program. Conversely, management must also provide funding for equipment and training for departments to realize these goals. As such, senior management must develop a facilitation and support plan for AM capital. Internal research and development budgets must be dedicated for AM to address technical gaps.

To encourage the adoption of AM by program and individuals, management must also consider negotiation and rewards. For example, a top-down-driven, reward-based programs can encourage employees to drive AM adoption in their business unit or department. For AM "nonbelievers," management may need to consider manipulation and co-optation to bring these individuals into successful AM teams and transition their thinking. This can also be accomplished by highlighting competitors' successes with AM. While this requires explicit and implicit coercion, it can be accomplished often times by showcasing how a lack of AM adoption will impact bottom-line business.

Business strategy tools

In addition, there are business strategy tools that are commonly used that can assist with establishing a sense of urgency and risk assessment. While it is important to identify opportunities for improvement, it is also critical to assess and mitigate risks. Risks can be threats and opportunities and

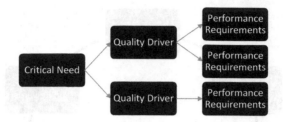

Figure 11.4 Tree diagram.

should be addressed through actions to avoid, mitigate, acknowledge, or accept these risks. Tools and techniques that can help identify and mitigate risks include decision trees, Strengths, Weaknesses, Opportunities, and Threats (SWOT) analysis, and risk grids.

Decision trees are a decision-making tool that illustrates decisions and their possible consequences. In particular, they are useful for clarifying choices, risks, objectives, monetary gains, and information needs. These diagrams are drawn from left to right using decision rules in which the outcome and the conditions form a conjunction using an "if" clause. Generally, the rule has the following form (as shown in Figure 11.4):

> *if* condition1 *and* condition2 *and* condition3 *then* outcome.

A SWOT analysis is a structured approach to evaluating a product, process, or service in terms of internal and external factors that can be favorable or unfavorable to their success. Strengths and opportunities (SO) ask "How can you use your strengths to take advantage of these opportunities?" Strengths and threats (ST) ask "How can you take advantage of your strengths to avoid real and potential threats?" Weaknesses and opportunities (WO) ask "How can you use your opportunities to overcome the weaknesses you are experiencing?" Weaknesses and threats (WT) ask "How can you minimize your weaknesses and avoid threats?" An example of a SWOT analysis is shown in Figure 11.5.

Next, let's apply the SWOT analysis approach to a typical AM challenge companies may face when first beginning their AM journey. Let's consider that we work for a small vehicle light manufacturing company. This company makes police, fire and ambulance lights for emergency responders and sells their product globally. The team wants to establish a sense of urgency to create a change in the company to implement AM into their company. The higher-level executives gather in a conference room and begin to perform a SWOT analysis on their product lines.

First, they start with their strengths. They determine that they have a strong understanding of polymer technology as applied in their light bar division. Through the years, they have become strong in understanding

Strengths	Weaknesses
What do you do well?	What could you improve?
What unique resources can you draw on?	Where do you have fewer resources than
What do others see as your strengths	others?
	What are others likely to see as
	weaknesses?
Intelligent resources – employment	Competitive salaries
Good intentions to want to do right things	Managers
Communication technologies	People believe their executives are not
Communication philosophies –	aligned – organizational direction
methodology	Leadership competency
Organizational direction	Learning gap
Strategic executives	
Opportunities	**Threats**
What opportunities are open to you?	What threats could harm you?
What trends could you take advantage of?	What is your competition doing?
How can you turn your strengths into	What threats do your weaknesses expose
opportunities?	you to?
Learning technologies to keep up with the	Keeping up with technology/new
times	learnings
Utilizing resources properly	Large generational gaps
Generational shifts	Creating work environments to engage all
Types of communication	generations and keep them attracted
Reverse mentoring	Retention
New revived energy	

Figure 11.5 SWOT analysis example.

how polymer materials have been designed and manufactured in lightbar products. They have a great understanding on how to manufacture these lightbars and the resulting quality you receive from them. How can they use AM to shore up these strengths?

Next, they look at their weaknesses. They know that they are weak in understanding how AM can help their company. They also are aware that they do not invest enough in research and development, which has led to their product lines becoming stagnant and becoming more of a commodity to their customers. There is not a lot of excitement from their customer base.

Next, they move on to their opportunities. They have heard that AM can be used as an opportunity to speed product development time. They also want to speed up the production line and they have heard of opportunities to use AM to build rapid tooling and fixtures.

Finally, the team assesses the threats. The team identifies that they do not have a great understanding of AM, and there are no subject matter experts within the company who can assist in developing the technology. Also, they have learned that the competing emergency warning lighting company also is using AM through marketing releases.

Now, that the team has conducted a thorough SWOT analysis they can start to define an action plan around AM technology that helps establish a sense of urgency. Using their strengths in polymer knowledge, they look to invest in an industrial scale polymer 3D printer. But because they are weak in understanding how they could use such machine, they concoct a plan to hire an AM consultant to help establish how AM can help them specifically. Based on the output on the company's opportunities and threats section, the consultant recommends resource planning for hiring an AM expert who can design rapid polymer tools that speed up the production lines. In the meantime, the consultant identifies parts that are polymer within their product lines that could potentially be AM'd and suggests ways to save part count and weight of their product through intelligent designs using AM that will ultimately become marketing content this company can release against their competition.

This entire exercise above helps establish a sense of urgency around the product lines within the emergency lighting company. In other words, because of our weaknesses and threats, if we do not do the following actions, our competitors will eat our lunch using AM.

After SWOT analysis, risks can be assessed using a risk assessment matrix (Figure 11.6). The risk impact assessment is a process of assessing the probabilities and consequences of risks if they occur. The results of a risk impact assessment will prioritize ranking from most to least critical with importance. Risks can be ranked in terms of their criticality to provide guidance where resources should be deployed or mitigation practices should be put in place in order to realize the risk events.

Coupling a risk grid with a SWOT analysis, executive leadership can gain a clear picture on the sense of urgency and how AM can address the urgency. Management must create a sense of urgency in order for AM

Probability		Low	Medium	High
	High	Low Impact High Probability 3	Medium Impact High Probability 2	High Impact High Probability 1
	Medium	Low Impact Medium Probability 4	Medium Impact Medium Probability 2	High Impact Medium Probability 2
	Low	Low Impact Low Probability 4	Medium Impact Low Probability 4	High Impact Low Probability 3
		Low	Medium	High
		Impact		

Figure 11.6 Risk grid.

introduction to gain momentum and be successful. Clear communication of the need to adopt AM is necessary to secure buy-in from all levels of the organization. Further, it is important for management to assess risk at this phase in order to develop plans to reduce or mitigate potential issues with AM adoption.

Reference

1. Kotter, J. (2012). *Leading Change*. Watertown, MA: Harvard Business Review Press.

chapter twelve

Create a guiding coalition

The previous chapter discussed how to create a sense of urgency with stakeholders to gain buy-in and cooperation to implement additive manufacturing (AM). Once the sense of urgency is established by senior managers, the focus is now on creating a guiding coalition, also commonly referred to as the change oversight team. This chapter discusses how to select and empower the coalition to lead the AM efforts throughout the organization.

Implementing AM in an organization is a significant effort that must be carefully managed and supported. Senior leadership cannot simply act on AM initiatives within the company alone. One must create a guiding coalition of multiple stakeholders to ensure proper success of implementing AM within the company. The coalition is a team of leaders and managers that are empowered by senior management to lead the change effort. Therefore, senior management must take time and care to select appropriate leaders within the organization that have technical knowledge of AM, understand the strategic vision of the organization, and have strong project management and interpersonal skills. In other words, senior management must select individuals with credibility, expertise, leadership, and position power that can work together. Simply put, if the coalition does not get along or does not have the technical knowledge, then AM efforts may fail.

You want a strong diverse mix of individuals involved in the AM effort and leaders must stay engaged once they announce a company-wide initiative such as AM. As an example, within a major company, an executive leader arbitrarily determined that the company would produce a new AM part a day. Why establish such a metric? Once the directive was established, the leader would stand back and wait for his/her team to report progress on a quarterly basis. Did the leader provide the resources for the team to achieve producing a new AM part a day? No. Was the leader engaged in the effort through active participation? Not really. The seemingly arbitrary decision to announce an initiative of producing one AM part a day fell flat. And once again, AM gets a credibility "black-eye." In short, no guiding coalition was formed for the effort and the technology suffered because of it.

AM insertion initiatives within companies also encounter the wrong people being placed in the wrong job role. The members of the coalition must have the AM technical skills to become credible within the organization. Would you trust a leader to lead an AM initiative if they have never visited and interacted regularly with the people in AM laboratory? Or, would you trust an AM leader who has never visited AM service bureau part suppliers? The AM industry is so new that finding an executive leader that has actually ran an AM machine at one point in their career is difficult, if not impossible. The point is that technical credibility must be attained within the skill mix of coalition members. Does every member of the team have had to run an AM machine before? No, but hopefully someone on your team has.

Simply forming the team also does not create a guiding coalition. Experts in any field, and AM is no exception, must have their egos checked by senior leaders. During idea generation sessions, throwing rocks, darts, or whatever colloquialism you want you use, should not be allowed by the team. If this becomes systemic behavior by one or more members of the team, they should immediately be removed. Though healthy debate is good to generate optimum ideas, constant obstructionists have no room in AM implementation collaboration sessions.

In building group dynamics for any new technology, organizations must have a good mix of people aligned to one common cause. For AM, the French and Raven's Five Forms of Power model is relevant [1]. Originally, they describe the five bases and in 1965, they added a sixth Form of Power. They are as follows:

1. *Legitimate*: This power given to the current AM leader that has the official job title. Basically, this is the person who bears the brunt of the responsibility of AM insertion. The "buck stops here guy/gal" is another way to describe this.
2. *Reward*: This power is held when one leader can offer compensation to another for compliance. So, in AM terms, this would be a person who has the ability to reward AM initiatives. It may or may not be the person listed above in #1.
3. *Expert*: This is the power wielded by the guru of AM. An expert in AM that understands the different technologies and how they may be applied. This may be an internal resource or initially an external consultant the company has hired to move AM forward.
4. *Referent*: This power is held by a seasoned person in the company that everyone respects. Perhaps not an AM expert, per se, but it could be a Chief Engineer of a program that can provide a vision of how the technology can be applied in large-scale form.
5. *Coercive*: This power type is often used for punishing people for not getting behind the AM bandwagon. This type of power would often

fall on first-line managers of AM group participants that can rein-
force an activity scope before it spirals out of control or even kick
AM obstructionists off the team.

6. *Informational*: This power results from the ability to manipulate
   who has access to what information to accomplish something. This
   can be a very important form of power. Having access to financial balance sheet statements, resource planning, and incentiviza-
   tion strategies are all sources of informational power. Within an
   AM coalition, this power would most likely be held by the senior
   executives.

Ideally, if you are serious about creating a guiding coalition of folks serious about pushing AM forward, you must consider a good mix of people
in the room that can wield each of the sources of power within the company. Selecting the right team is the first step in providing the foundation.
Senior leadership must also structure the coalition in such a way to ensure
its effectiveness. This means working outside of the normal hierarchical
structure. Instead, the coalition should report directly to senior leadership, which will help ensure decisions are best for the overall organization
during the change effort. In addition, off-site team building activities can
be effective in coordinating the efforts of the team and keeping the team
focused on the goals of AM adoption.

A key tool for creating the guiding coalition is Quality Function
Deployment (QFD), which translates the voice of the customer (VOC) into
the technical characteristics of a product. The VOC represents the desires
and requirements of the customer at all levels, translated into real terms
for consideration in the development of new products, services, and daily
business conduct. The VOC is a method for listening to and understanding the customer needs. It is important to have constant contact with the
customer to understand their changing needs.

It is critical to involve the customer(s) early in the implementation of
a new technology. The requirements must be based on customer needs to
ensure the final product is marketable. The requirements are then used to
determine the appropriate specifications necessary to meet the high-level
requirements.

At the beginning of AM implementation, it is essential for the team to
determine the customers and stakeholders. Customers typically include
peers, associates in your reporting line, supervisors, suppliers, other business areas, other groups in the organization, and external customers.

The team must identify primary and secondary stakeholders and
anyone affected by the introduction of AM. Once the stakeholders are
identified, the team needs to ensure they understand the stakeholders'
attitudes toward change and any potential reasons for resistance. As
part of this effort, the team should develop activities, plans, and actions

to overcome the resistance and barriers to change. To do this, the team should determine:

- How and when each stakeholder group should participate in the change
- How to gain buy in during the planning phase

Figure 12.1 provides a matrix to identify stakeholder groups, their role in AM adoption, and their impact and/or concerns.

Gathering the VOC is a critical part of introducing AM technology to ensure the desires and requirements of the customer are understood at all levels. The coalition can then translate the VOC into the development of new products, services, processes, and daily business conduct. The VOC cannot be gathered just at the beginning of the product development process because this only captures one point in time and customers' needs and desires change. It is important to have constant contact with the stakeholders and customers to understand their changing needs.

QFD is a systematic approach for translating consumer requirements into appropriate company requirements at each stage from research and product development to engineering and manufacturing to marketing/ sales and distribution. QFD makes use of the voice of the consumer throughout the process. The methods for QFD are simply to utilize consumer demands into designing quality functions and methods to achieve quality into subsystems and specific elements of processes. The basis for QFD is to take customer requirements from the VOC and relaying them into engineering terms to develop products or services. Graphs and matrices are utilized for QFD. A house type matrix is compiled to ensure the customer needs are being met into the transformation of the processes or services designed. QFD is a collection of matrices that are used to facilitate group discussions and decision-making. The QFD matrices are constructed using affinity diagrams, brainstorming, decision matrices, and tree diagrams.

Stakeholder/ Customer	Role Description	Impact/Concern	+/-

Figure 12.1 Customer/stakeholder matrix.

As a structured team approach, QFD can lead to a better understanding of customer requirements, which, in turn, lead to increased customer satisfaction and better teamwork. Also, understanding and integrating the VOC into the design reduces the time to market and development costs. QFD also enables a structured integration of competitive benchmarking in the design process. QFD also increases the team's ability to create innovate design solutions that provide enhanced capability and greater customer satisfaction. Finally, QFD provides documentation of the key design solutions.

There are seven key steps in creating the house of quality:

1. Determine what the customer wants. These are the "whats."
2. Determine how well the organization meets the customer requirements compared to the competition. This is the competitive assessment.
3. Determine where the focus should be to maximize return. This is based on the competitive assessment.
4. Develop methods to measure or control the product or process to ensure customer requirements are met. This is the "how."
5. Evaluate the proposed design requirements (hows) against the customer requirements (whats). This is performed in the relationship matrix.
6. Evaluate the design trade-offs. This is performed in correlations matrix or "roof" of the house of quality.
7. Determine the key design requirements that should be focused on. This is in the "basement" of the house of quality and is commonly referred to as the "how much."

A general overview of the House of Quality is provided in Figure 12.2.

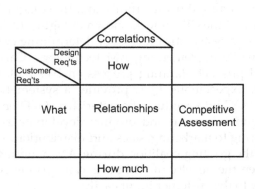

Figure 12.2 House of quality.

QFD is a structured methodology to identify and translate customer needs (wants) into technical requirements, measurable features, and characteristics. Proper execution of QFD forces the interaction with the product user to establish the VOC. QFD organizes the benefits of AM into a scoring system that is applied directly to the customer wants. The QFD works as a basic building block with a structured flow down process. Critical to Quality (CTQ) measures key characteristics of a product or process as defined by the customer. All CTQs are identified at one level, and the relating CTQs are built to the next level followed by the "hows." The "hows" at the next level become the "whats" at the next level. QFD follows a five-phase approach for AM:

- *Phase I*: Identify customer needs
  Customer requirements are identified and used to prioritize customer needs (CTQs)
  Construct the "House of Quality"
- *Phase II*: Functional requirements
  Use the prioritized customer needs to identify key functional requirements
  Determine high value offering concept
- *Phase III*: Offering design
  Functional requirements are used to systematically identify design requirements
- *Phase IV*: Part design
  Product design requirements are used to identify critical part characteristics
- *Phase V*: Process design
  Specific process requirements are identified to ensure key part requirements from Phase IV are controlled and maintained

The House of Quality is the initial QFD matrix. It is where the customer requirements are analyzed using affinity and tree diagrams. The graphical representation of the QFD flow is shown in Figure 12.3.

QFD provides a systematic process for determining critical-to-quality characteristics while implementing AM technology. Carrying the VOC throughout AM product planning process enables a clear understanding of customer expectations. QFD provides a systematic approach for translating the VOC into product characteristics. QFD ensures the VOC is heard throughout research and product development to engineering and manufacturing to marketing/sales and distribution. This is essential for supporting the guiding coalition during AM adoption, especially in AM design when the novelty of AM parts can become customized and tailored directly to the customer input easily.

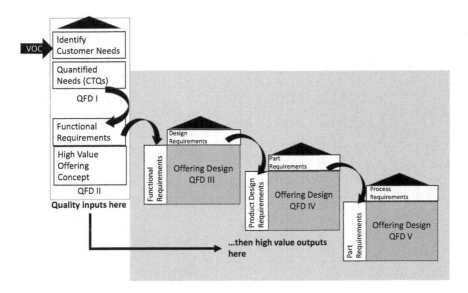

Figure 12.3 QFD in AM adoption.

While the coalition is given responsibility for the success of AM implementation, it is important that senior leadership remains actively engaged throughout. In order for stakeholders throughout the organization to buy in to the AM adoption, senior leadership must also show their support through engagement. This will alleviate possible resistance to change from stakeholders. Through proper selection and empowerment, the coalition can increase the likelihood of successful AM adoption. In addition, senior management must stay involved throughout the change effort.

Reference

1. French, J. R. P., Raven, B. (1959). The bases of social power. In: D. Cartwright and A. Zander. *Group Dynamics*. New York: Harper & Row.

chapter thirteen

Develop a vision and strategy

Organizationally, many of the challenges with inserting additive manu-facturing (AM) into the culture stems from inherent dysfunction within the organization prior to AM insertion. This is where a key focus on the vision and strategy for AM adoption is critical.

Once senior leadership has instilled a sense of urgency and created a guiding coalition, the next step is to develop a change vision and strategy. Developing the AM change vision and strategy takes time and should not be rushed. It should clearly represent the ideal future state after AM adoption as this helps motivate and gain buy-in from all levels of the orga-nization. In addition, the vision will help guide decisions that enable AM adoption.

For individuals to accept the AM change vision and strategy, it must be feasible and desirable. The strategy should provide sufficient infor-mation for individuals to see that the journey is realistic and achievable. Further, employees must clearly see that the future for the organization and their personal growth is better with AM. These are both necessary to create a compelling reason for employees to embrace the change.

Senior leadership is ultimately responsible for setting the organiza-tion's direction. While the guiding coalition oversees much of the AM implementation, they should not be responsible for creating the vision and strategy. However, once a draft is complete, the guiding coalition can offer revision suggestions to ensure successful AM adoption. The vision and strategy should be developed using market data, emerging business trends, and company business data, among other data sources.

A successful AM vision and strategy should be easy to communicate and should not be complex. It should include areas of change, achievable targets, measures of success, and long-term interests of organizational stakeholders. It is necessary for all employees and managers to under-stand the organization's direction. If they do not, AM adoption will not be successful.

No organization is perfect, but most should adhere to Beckhard's goals, roles, processes, and interpersonal relationships (GRPI) model for a basic structure as shown in Figure 13.1. You might be surprised how often this model is not followed within companies.

In adopting any major initiative, there will be obstacles that hinder change and reasons that support change. Lewin's force field analysis is

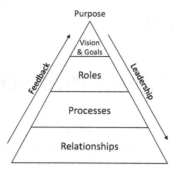

Figure 13.1 Beckhard's GRPI model [1].

one method for identifying these areas. During AM adoption, there will be aspects that help the adoption, these are driving forces. Conversely, there are aspects that will prevent AM adoption from occurring, these are restraining forces. By identifying these prior to AM adoption, senior leadership can take appropriate actions. Figure 13.2 provides a force field analysis for slow AM adoption.

Dana Larsen, a Senior Director of Organization Development, is often asked by executives to "fix" organizations that are not functioning well. He states that leaders like to prescribe the solution to the problem by saying the teams need more "team building" activities to make everyone get along. Perhaps team bowling nights or paintball are in order?

However, through his years of experience, he goes back to Figure 13.1 and through a series of interviews, he first looks at how well the leaders

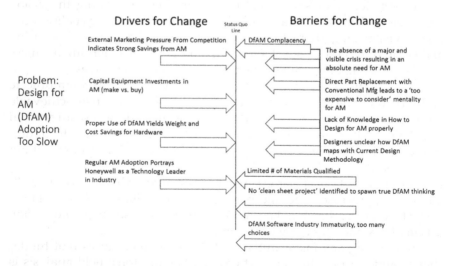

Figure 13.2 AM adoption force field analysis.

have defined a vision and respective goals for the organization. In some cases, there may be no vision on what the team should do, nor any defined goal associated with the team. This is a foundational element to a proper working team. If there is a lack of vision and defined goals within an organizational team, then establishing an organizational charter may need to be order to move forward. This should be a fairly simple exercise to accomplish. Each charter should include:

- Mission of the organization in a single sentence
- Vision of where the organization should be 3–5 years out
- Overall goals of the team based on the mission and vision above
- Activities defined that align to the goals
- Operating metrics defining success for this organization
- Interdependencies defined with respect to other functions
- Risks that may be real to the organization that may prevent the mission stated above from occurring

Next, if a vision and goals are firmly established, Dana looks to see if roles and responsibilities are clearly defined within the team. This may also be a huge problem. Perhaps no one took the time to clearly associate the appropriate tasks with each of the team members. If roles and responsibilities are not well established, perhaps a RACI (responsible, accountable, consult, and inform) or responsibilities matrix is a good place to start to clearly define who is doing what on the team. It is simple, yet powerful tool for assigning cross-functional responsibilities. To setup a responsibility matrix, a list of tasks are compiled and R's, A's, C's, and I's are placed next to individuals names that represent all the stakeholders in an AM team. R's stand for the person who is responsible for the action; in other words, the person who actually accomplishes the task. The A's are assigned to the people who are ultimately accountable for the effort, such as a first level manager. C's are defined as people who are consultants to the role. These are typically subject matter experts in the field. I's are listed next to people who need to be informed of tasks. Each task should have one individual who is accountable, while there may be shared responsibility in some cases. An example responsibility matrix is provided in Figure 13.3 for AM.

Department	Tasks	People											
		Frank	Tom	Jason	Kirsten	Rebecca	Justin	Bryan	Jeff	Amir	Seth	Fred	Allen
Lab Operations	Material order/re-order	I	R	I							A	I	
	Operating of the SLM machine	R	I	C							A	I	
	Operating the EOS machine	C	I	R							A	I	
	Pre-processing the AM builds	I		I			R	C			A	I	
	Post-processing the AM builds				I	R		I			A	I	I
Design	CAD file design and transfer	C			R					C	I	I	A
	Distortion modeling analysis							R		I	I	A	
Quality	White light inspection				I	I			I	R			A
	Testing of witness coupons				I	I			R				A

Figure 13.3 Example RACI matrix.

Back to Figure 13.1, if roles are clearly defined, next, Dana moves on to look at how systems and processes are setup for the team members. In many cases, processes are not clearly documented, computing systems make conducting work difficult, bad processes lead to redundant work environments, etc. Given that industry consists of many systems all working in parallel to drive value into the company. One may look to understand these processes by physically walking or observing the process in action, auditing the results then illustrating the process and using basic process flow diagramming or value stream mapping principles. If processes are found not to be under control, one may consider utilizing standard work practices to get the process flow consistent. When people depend on each other, standard work can make any work process easier, more predictable, and less prone to error. Standardized work may consist of:

- Job breakdown worksheets defining job tasks in great detail
- Process checklists
- Forms and procedures
- 5S (sort, set in order, shine, standardize, and sustain)

Based on many years' experience in this type of role, Dana's theory is that in most cases, if there are relationship issues with teams, it is merely a symptom of something wrong in defining either the vision or roles or processes not working. In other words, it generally isn't the team's fault as much as it is the executive's fault for not setting up the organization well in the first place. Dana admits this is a tough message for him to deliver to the leadership that hired him to "fix" the organization. He believes that if the top three elements of the pyramid are well established, the relationship base of the pyramid will naturally be improved.

The point of looking at the model in Figure 13.1 is to understand how an organization should perform. The authors coin the term, "organizational fitness" to describe the effectiveness of a working organization. Is your organization team out of shape or obese, or is it running in a lean and efficient way? Simply inserting AM technology in your organization with broken goals, role confusion, and dysfunctional processes will only exacerbate relationship issues and personality conflicts will emerge within your AM teams.

In developing the AM vision and strategy, senior leadership should also have a clear project plan that is communicated. Project management, which relies on knowledge thus acquired, is the pursuit of organizational goals within the constraints of time, cost, and quality expectations. This can be summarized by a few basic questions. They are as follows:

- What needs to be done for AM implementation?
- What can realistically be done for AM implementation?

- What will be done for AM implementation?
- Who will do it?
- When will it be done?
- Where will it be done?
- How will it be done?

The factors of time, cost, and quality are synchronized to answer the above questions and must be managed and controlled within the constraints of the iron triangle depicted in Figure 13.4.

In this case, quality represents the composite collection of AM requirements. Project lead time is most always improved by choosing an AM solution over a conventional manufacturing process. However, depending on the AM solution, costs may be higher. In a situation where precise optimization is not possible, trade-offs among the three factors of success are required. A rigid (iron) triangle of constraints encases the project (Figure 13.4). Everything must be accomplished within the boundaries of time, cost, and quality. If better quality is expected, a compromise along the axes of time and cost must be executed, thereby altering the shape of the triangle.

The trade-off relationships are not linear and must be visualized in a multidimensional context. This is better articulated by a three-dimensional (3-D) view of the systems constraints as shown in Figure 13.5. Scope requirements of an AM project determine the project boundary and trade-offs must be achieved within that boundary. If we label the eight corners of the box as (a), (b), (c) ... (h), we can iteratively assess the

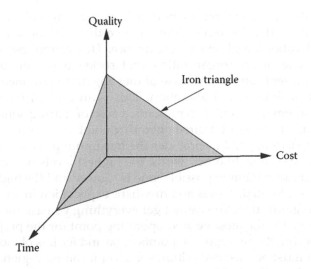

Figure 13.4 Project constraints of cost, time, and quality.

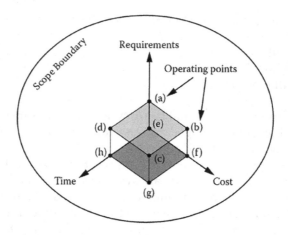

Figure 13.5 Systems constraints of cost, time, and quality within iron triangle.

best operating point for a project. For example, we can address the following two operational questions:

1. From the point of view of the AM project sponsor, which corner is the most desired operating point in terms of combination of requirements, time, and cost?
2. From the point of view of the AM project executor, which corner is the most desired operating point in terms of combination of requirements, time, and cost?

Note that all the corners represent extreme operating points. We notice that point (e) is the do-nothing state, where there are no requirements, no time allocations, and no cost incurrence. This cannot be the desired operating state of any organization that seeks to remain productive. Point (a) represents an extreme case of meeting all requirements with no investment of time or cost allocation. This is an unrealistic extreme in any practical environment. It represents a case of getting something for nothing. Yet, it is the most desired operating point for the project sponsor.

By comparison, point (c) provides the maximum possible for requirements, cost, and time. In other words, the highest levels of requirements can be met if the maximum possible time is allowed and the highest possible budget is allocated. This is an unrealistic expectation in any resource-conscious organization. You cannot get everything you ask for to execute a project. Yet, it is the most desired operating point for the project executor. Considering the two extreme points of (a) and (c), it is obvious that an AM project must be executed within some compromise region inside the scope boundary. Figure 13.6 shows a possible compromise surface with

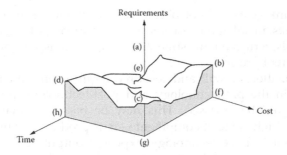

Figure 13.6 Compromise surface for cost, time, and requirements trade-off.

peaks and valleys representing give-and-take trade-off points within the constrained box. With proper control strategies, the project team can guide the project in the appropriate directions. The challenge is to devise an analytical modeling technique to guide decision-making over the compromise region. If we could collect sets of data over several repetitions of identical projects, we could model a decision surface to guide future executions of similar projects. In an organization that has poor organizational fitness, you may notice a constant changing of requirements, funding, and scope during the course of the AM project.

Justification should be compiled when making an investment of any kind toward a project. This entails the comparison of industry objectives with the overall organizational goals in areas such as target market, growth potential, sustainability goals, profitability, capital investment, skill and resource requirements, risk exposure, and national sustainability improvement.

The target market includes the customers of the AM technology. This should also address items such as market cost of the proposed product, assessment of competition, and market share. Growth potential for AM addresses the short-range expectations, long-range expectations, future competitiveness, future capability, and prevailing size and strength of the competition.

AM technology should also be evaluated in terms of direct and indirect benefits toward the organizations sustainability goals. These may include product price versus value, increase in international trade, improved standard of living, cleaner environment, safer workplace, and improved productivity.

An analysis of how AM technology will contribute to profitability should consider past performance of the technology, incremental benefits of the new technology versus conventional technology, and value added by the new technology. In addition, a comprehensive economic analysis should play a significant role in the technology assessment process, as capital investments may be significant. This may cover an evaluation of fixed and sunk costs, cost of obsolescence, maintenance requirements, recurring costs, installation cost, space requirement cost, capital substitution potentials, return on investment, tax implications, cost of capital, and other concurrent projects.

The utilization of skill and resource requirements (manpower and equipment) in the pre-technology and post-technology phases of AM adoption should be assessed. This may be based on material input/output flows, high value of equipment versus productivity improvement, required inputs for the technology, expected output of the technology, and utilization of technical and nontechnical personnel.

Uncertainty is a reality in any technology adoption efforts and AM adoption is no exception. Risk exposure should be assessed for the initial investment, return on investment, payback period, public reactions, environmental impact, and volatility of the technology.

Finally, an analysis of how AM technology may contribute to national sustainability goals may be verified by studying industry throughput, efficiency of production processes, utilization of raw materials, equipment maintenance, absenteeism, learning rate, and design-to-production cycle.

A clear AM vision and strategy greatly enhances the chances of AM adoption success. Senior leadership, with the help of the guiding coalition, must communicate the rationale and need for change. It must be simple and straightforward in order to gain buy-in from employees and manager. This will drive the change in the organizational culture and pave the way for AM adoption.

Reference

1. Beckhard, R. (1972). Optimizing team building efforts. *Journal of Contemporary Business*, 3, 23–27.

chapter fourteen

Communicate the change vision

Communicating the change effectively within organizations is absolutely critical in successful implementation of additive manufacturing (AM). A communications plan should be established that clearly defines which groups within the organization shall be informed and why they should be informed. There will be different stages the workforce goes through to absorb the news of the impending change. How your team's emotional response and expectations are managed as they move through the different stages is critical.

Halford E. Luccock once said, "No one can whistle a symphony. It takes an orchestra to play it." Once an AM strategy is agreed upon by all the stakeholders and funding is allocated to begin AM projects within the organization, a clear communication of that strategy should be the next step for leadership. This includes a clear direction and purpose of what is happening along with focused goals and rationale for the changes being made. It is taking what will be a complex endeavor and make it simple. While there is considerable data and metrics that can be shared as the business case for change, these should not be the core message of the change vision communication of implementing AM. Rather, it should be simple, concise, and appealing. When this message is communicated effectively and the guiding coalition promotes organizational understanding, commitment is gained from employees and managers. Effective and frequent communication drives change.

To see exactly how important this step is, let's consider a scenario that includes an industrial company that establishes an urgency to use AM, they create a guiding coalition and vision and strategy, but the leadership becomes so excited that it starts to procure AM equipment. In their excitement, they neglect to inform the company of the decision to acquire these assets. Within a few months, the machines begin to show up on the facility loading dock ready for installation at the company. People are largely unaware of the recent purchase because the management team forgot to communicate the change vision to the company. As a result, people begin to start rumors. People begin to whisper, what are these machines for? Are they going to make my job obsolete? Who will become trained on the technology? Are they hiring people from outside to run them? I don't want to lose my job, how do I become trained? Next, the speculation begins around the questions. I bet management is here to replace our jobs with those machines. They've been wanting to modernize our livelihood for some time! Like the proverbial telephone game, without a clear communication plan, the simple acquisition of AM equipment to improve company product offerings, can create a union grievance against management by accusations from workers about replacing people's jobs. This is only an example of how a communication plan is desperately needed to make people feel more comfortable with the impending change of adding AM to the company culture.

There are several required elements for change:

- *Change objectives*: Clearly define the objectives targeted through your AM transition.
- *Change vision*: Establish the vision for the AM change and communicate it clearly throughout the entire organization.
- *Champion, sponsors, facilitators, teams*: Identify the people who will lead the change, fund the change, and execute the charge.
- *Training*: Establish a clear training plan for AM (covered in chapter fifteen).
- *Systems that enable and support AM*: Ensure that the right software tools, capital equipment, post-processing equipment, and your enterprise resource planning system are supporting your AM efforts.
- *AM performance metric(s)*: Employees will work to their metrics. Be smart about what metrics are set up for your AM initiatives.

Change management

People naturally fear change. They are afraid of change because it makes comfortable ways uncomfortable while changing the normal way of operating. Employees may fear because they may feel they are losing ownership. While others may fear the unknown, since they are being asked to take on new tasks and responsibilities outside of their comfort zone. Effectively implementing change involves frequently incorporating new competencies. New competencies will enable further education and training for all employees. Once this new mentality becomes the norm, the change is viewed as good. In the case of AM, change involves implementing what may seem very daunting – a technologically advanced manufacturing process. Change, and its impact on employees, must be considered when adopting AM.

Heller [1] developed a model for managing change, which is shown in Figure 14.1. The response to change is based on how individuals handle the change. Individuals will go through each type of response, although some individuals may take longer in the different response stages. By developing a communication plan and effectively communicating the

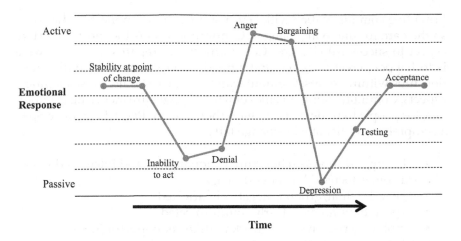

Figure 14.1 Change management model.

need for change and how it will affect individuals, people can move quicker to acceptance.

Communicating change

Effective communication should be frequent and simple. This fosters understanding. If the message is too complex, it will be difficult to gain buy-in. Managers and employees cannot embrace what they cannot understand. Jargon and buzzwords should also be avoided. People will question if it is a "flavor of the month" or a true organizational change.

It is also important to communicate the change vision through multiple means – e-mail, group meetings, town hall meetings, informal meetings, written memos, newsletters, company intranet, etc. The communication must be consistent and frequent. Not all employees and managers check their e-mails frequently or read them thoroughly. The same goes for newsletters and memos. Therefore, it is important to spread the message through as many possible avenues as possible to reach everyone in the company.

While verbal communication is important, people often retain more through pictures or diagrams. Particularly with a new technology such as AM, images may help employees grasp the concept and the changes quicker. These images can also provide a compelling picture of the need for change and how it can benefit the customer.

Utilizing AM technology can help the business grow by being more efficient and innovative. The systems being used help the business grow and the benefits of the new technologies attract new customers. To alleviate fears throughout the organization, a communication plan is essential.

Providing communication with key customers and suppliers is important so they are in tune with changes. Communication is the most important aspect to successful changes. Communication prepares others for what is to happen, creates shared and agreed upon visions, and builds relationships within groups. Communicating change should consist of many aspects. Communicating specifics of what changes are incurring is important for people to understand to gain their buy in. Recardo [2] developed a comprehensive list of why change fails:

- The organization's architecture is not aligned and integrated with a customer-focused business strategy
- The individual/group is negatively affected
- Expectations are not clearly communicated
- Employees perceive more work with fewer opportunities
- Change requires altering a long-standing habit
- Relationships harbor unresolved past resentments
- Employees fear job security
- No accountability or ownership of results
- Poor internal communications
- Insufficient resource allocation
- It was poorly introduced
- Inadequate monitoring and assessment of change progress
- Dis-empowerment
- Sabotage

These fears should be addressed in the communication plan. People need to understand the importance of change to the business and the reasoning behind it. It is important for the person delivering that message to understand themselves the reason of the change. The downfalls of what could happen if the change does not occur should be explained at this point in time as well. Explaining the urgency of the change to occur will speed up progress. Specifics on timelines and milestones also help drive the urgency.

Motivating people to change is imperative, so they feel empowered to change and do not just make changes that will not be able to be sustained. Further, understanding the organization's need for change is imperative in order to embrace the change. This helps the individuals to also understand "what's in it for me?", which is essential for buy-in.

Senior leadership and the guiding coalition should also publicize the wins of work well done or employees that have embraced change. Preparing information and updates on results, changes, and progress will help people visualize the impact the changes are making. It is important to communicate as many of these wins in person. Many people will begin asking questions; admitting not knowing information is acceptable as long as there are means to showing you will find out the information.

Being truthful and transparent will help expectations more clear even when discussing the downfalls.

Re-capping of past communications and topics with outcomes will help gain credibility on the work that was performed along with the successes that can be seen.

Many people will want to understand their roles and responsibilities; showing favoritism should not occur. Consistency is a key for success when dealing with different people and different personalities. Mistakes will happen, so learning from mistakes while being transparent about why mistakes incurred will prevent them from happening in the future. Capturing best practices and learning techniques will help benchmark ways for working in the future.

If you are looking to push initiatives within the company from solely a downward communication strategy, there are pitfalls in this approach. Filtering of the AM change messaging can lead to messaging information loss as shown in Figure 14.2.

Conversely, engagement from all levels of employees will help the support of projects and increase awareness of new AM initiatives.

Successful AM implementations come from asking the basic questions and blending them appropriate for the proper answers (who, what, where, why, when, and how). How to change and what to change are the big factors that will explain the success of change. Once the establishments of these measures are in place, the implementation process is simplified. Successful workplace changes occur with the proper foundation and framework that has been established through management and employees through a shared vision. People must drive these visions together with their different skill sets. The cross-functional groups working together as teams are what drive successful change. The real change comes from the people combined together.

Figure 14.2 Filtering of downward communication.

Direct, frequent, and consistent communication also gives visibility on the changes happening with less ambiguity on why the decisions were made to adopt AM. Communication must also be multidirectional and not just top down. The open exchange of ideas must be able to occur in order to change. If employees are afraid to bring up their ideas on AM, they will continue to resist change because they are not being heard. It is essential to gain employees' trust and allow them to speak their minds. Once they describe the feelings they are having and the reasoning behind them, it is important to understand the position they are in. By minimizing the fears your employees have from AM adoption, you will slowly gain their trust and alleviate fears. Knowing what is driving the change and how it will affect individuals and the organization is an important step to acceptance of change.

Information sharing during the AM implementation is a powerful means for communication. Information given to one another builds teams to share known information and makes strategic decisions from the data. A company's culture is based on this information sharing. Culture is based on beliefs and values, and the combination and agreement of the two. Wanting to increase productivity and efficiency is part of the culture. If all people believe in this, the vision is the same. If some people have not bought in to the vision or need to change, then a culture shift needs to take place. Once the culture has the same enterprises of beliefs, the overall vision can be achieved by all entities working on the same goals due to their beliefs.

Finally, senior leadership and the guiding coalition must "walk the talk," If inconsistencies arise between leadership behavior and the change vision, they must be handled immediately. Those employees and managers that are on the fence or resistant to change will notice these inconsistencies. It will affect their motivation and potential buy-in. Leaders must respond with honest and direct explanations and provide plans on how to move forward. Care and attention must be made by senior leadership and the guiding coalition to support the change vision in their actions.

By effectively communicating the change vision in simple terms through multiple avenues, senior leadership and the guiding coalition are more likely to gain commitment. Communication must be two-way and consistent with leadership actions. Finally, the message must be clear through images and words. Once the change vision is clear, leadership can focus on AM implementation.

References

1. Heller, R. (2009). *Managing Change*. Dorling Kindersley Ltd. London, England.
2. Recardo, R. (1995). *Overcoming Resistance to Change*. National Productivity Review. John Wiley & Sons, Inc. Hoboken, New Jersey.

chapter fifteen

Empower broad-based action

Once the senior leadership has created a sense of urgency, identified a guiding coalition, and developed a change vision to drive change, the focus turns to additive manufacturing (AM) adoption by empowering broad-based action. This stage is the implementation of AM within operations and design and is where senior leadership and the guiding coalition must remove obstacles that do not support the change vision. While developing the change vision is difficult, now leadership must make the change happen, which is often the most difficult part.

As one of the critical aspects of change management, empowering broad-based action is a call to arms for the organization. This will include removing as many obstacles as possible in the path of AM deployment. As with any corporate improvement activity, with AM being no different, an organization must cement the new initiative broadly throughout the company. "With this effort, organizational processes, structures, procedures, and reward systems will need to be aligned with the new change vision as necessary. Empowering broad based action also involves investment in employee and managerial training and development." This is a necessary expense, not a luxury expense (https://managementisajourney.com/leading-change-step-5-empower-broad-based-action/).

So, how does this vision apply to AM? First, organizational processes must be set in place to adopt this new manufacturing method. For example, all enterprise resource planning (ERP) systems an organization uses must be adapted for AM. Electronic planning systems, routers, quality documentation, purchasing, receiving forms; basically, all digital and paper forms that are necessary to manufacture a certified product must be changed.

But fixing ERP systems and processes are fairly straightforward to resolve. There lies a more sinister culprit to prevent broad-based AM action, the troublesome supervisor. "Often these managers have dozens of interrelated habits that add up to a style of management that inhibits change. They may not actively undermine the effort, but they are simply not wired to go along with what the change requires.

Often enthusiastic change agents refuse to confront these people. Easy solutions to this problem don't exist. Sometimes managers will concoct elaborate strategies or attempt manipulation to deal with these people. If done skillfully, this only slows the process and, if exposed, looks terrible – sleazy, cruel and unfair – and undermines the entire effort. Typically, the best solution is honest dialogue."[1] Unofficial AM leaders can be identified by their ability to influence others and engage in activities. It is to a manager's benefit to utilize these leaders to help the change. Assessing the current structure of whom the natural leaders are, whom the resistance comes from, and who the followers may be will make the process easier to understand and manage. Simple questions can be asked such as:

- Who do people listen to?
- Who will adapt to change easily?
- Who will resist change?
- Who will help with the alignment of change?
- Who will motivate others?

In this step, it's also important to understand why people resist change. Resistance occurs at every level of the organization. Change cannot be forced on people. Leading and implementing change requires working with individuals and bringing them along. Senior leadership and the guiding coalition must have people skills. They must understand that resistance is logical to that person and it makes sense to them. The resistance is from their perspective and you must put yourself in their shoes. If people believe that a change is in their best interest, they will not resist the change. Chapter fourteen discussed the reasons people resist change. Those reasons must be addressed before you can empower change.

Also, reward systems must be designed to promote the use of AM within the organization. Consider creating a design challenge competition within your company that rewards the clever application of AM design. Consider how to metricize the use of the technology. Should you reward employees with how many AM ideas they create? Probably not, this will only lead to a large amount of subpar ideas that an AM resource must pick through to find a suitable AM idea.

Alternatively, organizations should contemplate rewarding the proposed "value" of the AM idea submitted, not the "quantity" of AM ideas submitted. What type of incentive should be offered? In practice, I've seen company cash awards, parking spaces, and time off with pay used effectively in these type of reward systems.

As with any new technology, investment in AM training at the employee and managerial level is absolutely critical. This cannot be stated enough. In fact, let me repeat this … investment in AM training at the managerial and employee level is *CRITICAL, it is not an option!* Too many times have AM experts been brought into a company to "change" how a company uses AM only to discover the masses of design engineers,

[1] www.cfma.org/content.cfm?ItemNumber=2378.

manufacturing engineers, project managers, and functional managers completely clueless about AM technology in general, let alone how to deploy it within a company. A formal training plan must be in place and executed at the engineer level, the project management level, the management level, the sales level, and the manufacturing level with each function specialized in the curriculum.

The approach above may sound extreme, but it is not, it is necessary to empower broad-based action within your respective company. Most companies serious about AM technology recognize the education shortfall in AM and have already implemented isolated areas of training. AM training consultants have virtually sprung up out of nowhere and will be expected to grow in this field. However, be cautious when selecting consultants to assist in educating your employees by being wary of the "charlatans" of AM and separate them from the people with real-world experience driving home AM initiatives. Ask questions of each AM training consultant evaluated. Effective questions might include: (1) How many years of experience do you have in AM technology? (2) What companies have you worked for? (3) How many AM parts have you designed/built/ implemented within each company? (4) Which specific AM technologies are you intimately familiar with?

Though not an endorsement, at the time of writing, some examples of qualified consultant groups providing training in this area includes, The Barnes Group Advisors, MIT Professional Education, UL Additive Manufacturing Competency Center, and SAE International. That said, the quality level within these groups tends to vary widely based on the level of experience and industrial background so get engage to ensure your team is getting useful support. Each group varies in the format of training which can include on-site work sessions held at your company, web-based learning, and off-site lectures held at the training consultancy's site.

As you go down this journey to AM education, one of the first steps includes the notional sketching of an AM training plan. Providing training enables employees to support the change vision. It is also in the best interest of managers, employees, and the organization as a whole. When identifying training opportunities, the senior leadership team and the guiding coalition should consider the not only the new skills needed, but also the attitudes and behaviors. This will enable the appropriate training and development efforts to be designed and/or adapted to align with what is needed for AM efforts. Adapting AM content from the University of Minnesota Human Resource Management model,[2] the basic outline of a comprehensive training plan includes the following:

---

[2] http://open.lib.umn.edu/humanresourcemanagement/chapter/8-4-designing-a-training-program/.

1. *Needs assessment*: This will include which groups within the company needs specific AM training. Ideally, map out each function that will "touch" AM technology once fully deployed within the company. This should also respect changing demographics within the company's workforce and AM technology trends.
2. *Learning objectives*: A learning objective is what you want the learner to be able to do, explain, or demonstrate at the end of the training period. Good learning objectives are performance based and clear, and the end result of the learning objective can be observable or measured in some way. An AM example would be "Want AM designers to be able to recognize which designs should or should not be AM." "Want hiring managers to pick out resume keywords for AM talent acquisition." "Need manufacturing engineers to be able to prepare an AM build job for printing." Desire for a materials engineer to learn the latest mechanical property information related to existing and new AM materials development and how to best generate this type of data." Note that the learning objectives above change based on which audience is receiving the AM training.
3. *Learning style*: Define which type of learning style works best for absorbing information regarding AM. Typically, there are three different learning styles that students utilize during their training: visual (learning through observation), auditory (learns by listening to lecture), and kinesthetic (learns by feeling and actually touching things). A well-designed AM training program will include all three types of learning style instruction to support this variation for the target audience.
4. *Delivery mode*: The delivery format of the training will accommodate the learning style defined above. Web-based, job-shadowing, lecture, AM equipment interfacing, etc. I've found that the best delivery format for AM training generally involved hands on laboratory access to accommodate the kinesthetic learners, formal lecture for the auditory learners with a good amount of case studies for the visual learners. If the class is held in a room, what type of room should it be? A conference room with lots of white boards or perhaps a dedicated training room with an AM display case showing case examples of AM parts?
5. *Budget*: How much should/can you pay consultants to teach the course? How many folks do you want trained at a time? How much is your internal AM subject matter expertise worth from an opportunity cost standpoint? These are all questions that need answers to define the budget constraints.
6. *Delivery style*: Blasting through PowerPoint presentations for hours on end is not effective. Breakout discussion, ice breakers, working groups, hosting panel Q&A discussions are all options. Define what

the delivery style should be upfront with the objective to encourage as much discussion and questions as possible.

7. *Audience consideration*: Since AM touches so many departments cross-functionally, should you have one cross functional class of participants, or specialize and have very specific content delivered per each functional group requirements? Do you want to only train new employees, seasoned employees, or both?

8. *Content development*: Should this training be developed in-house with company specific content, or rely on out of house consultants to develop? Or, perhaps a hybrid mix of both approaches. This direction will link directly to the development of learning objectives above.

9. *Timelines*: Should the AM training be mandatory before the employee starts with the company, or as an optional requirement that is encouraged to be completed by a specific time? Related to #5 Budget above, how much time do you want the participants to spend in training?

10. *Communication of training*: With the training now conceived, how will the message of the new AM training opportunity be rolled out to the organization? Will this be messaging through a mass departmental e-mail, a poster in the break room or can it be posted on the company intranet? In my experience, it is amazing how many people were unaware that training in AM was being conducted in the same company that they work in. A strong communication plan should be developed and acted out in order to maximize the demand for the class. Be sure to engage with key managers to identify specific employees for these initial training opportunities. They will appreciate the consideration and ability to support the professional development of their most valuable team members and you are likely to get higher performing candidates.

11. *Measuring effectiveness*: One of the most often overlooked elements of training is the self-assessment of the training. Ideally, you should ensure that the training met the learning outcomes you determined up front, what benefit the organization will receive from the training, and how did the students feel regarding the training. Should this be in the form of an anonymous survey? If so, who will write the survey? What specific questions should be on the survey? Perhaps a quiz at the end of the AM class that results in a certificate? What are the next steps in the overall training program and do we need to work with human resources to implement a competency model?

Once the AM training plan becomes formalized, then the company must figure out how to stage the training modules release within company and mix in the right participants based on project schedules and targeted AM applications, as notionally shown in Figure 15.1.

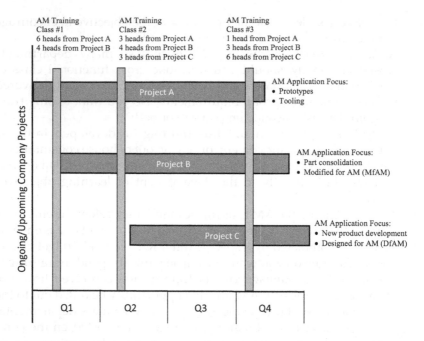

AM Training
Class #1
6 heads from Project A
4 heads from Project B

AM Training
Class #2
3 heads from Project A
4 heads from Project B
3 heads from Project C

AM Training
Class #3
1 head from Project A
3 heads from Project B
6 heads from Project C

Project A

AM Application Focus:
• Prototypes
• Tooling

Project B

AM Application Focus:
• Part consolidation
• Modified for AM (MfAM)

Project C

AM Application Focus:
• New product development
• Designed for AM (DfAM)

Ongoing/Upcoming Company Projects

Q1   Q2   Q3   Q4

Figure 15.1 AM training deployment plan coincides with company project.

Once metrics are recorded on the number of students trained in AM technology, next, the company should tie the investment of training back to tangible $'s saved for the company as notionally shown in Figure 15.2. It is important to note that the vertical axis can be whatever performance indicator your company wants to use to metricize AM potential, just ensure that you have a consistent method to measure your AM project value within the company. This step is critical to document the return on the training investment that must be made by the company. If positive returns on the training investment are realized, then it becomes easier to justify an expansion of AM training to other groups within the company and even expanding AM training into your company's supply chain.

Empowering change efforts within the organization is critical for AM adoption. Failure to empower employees and managers can lead to stalled efforts at the beginning of implementation. In addition, it is easy for organizations to maintain status quo if focus is not placed on taking action and driving change. The combination of skill sets necessary for AM adoption will help with challenging the status quo. Combining the ideas from stakeholder to come to a uniform decision will help the group to find strategic opportunities. Leadership also needs to be flexible during the AM implementation, and one individual cannot overpower the conversation or ideas being given. Adaptability of other people's viewpoints,

Figure 15.2 Relationship between employees trained in AM and value.

schedules, concerns, and styles should be taken into consideration and openly discussed. Leadership must show that they are open to change and suggestions. This will help others within the organization appreciate your values and anticipate the change. Change also takes time and that time it takes to reach acceptance is different for everyone.

Leadership, which is often focused on short-term gains for shareholders, must show they understand that AM adoption takes time and give the employees time and attention they need to implement the technology effectively. Further, people need reassurance during changing times, and as a leader that reassurance and commitment needs to be provided and felt. Leaders' actions and words must resonate the organization's values and be aligned with the change vision. Leaders need to show that they are changing the status quo for the benefit of the project and the business instead of being seen from the perspective of simply addressing the latest technology fad and promoting their own career growth.

chapter sixteen

Generate short-term wins

In order to drive momentum with the adoption of any new technology, management should focus on short term wins. With respect to additive manufacturing (AM), the focus should be on using the new technology for simple, cost-effective parts prior to focusing on complex, expensive parts. Through successful work on the low hanging fruit, employees will feel more confidents as they move to more complex products. This transition also enables management to ramp up training throughout the facility for AM practices. An approach for product selection is the Pugh concept selection matrix.

Adopting AM is a significant undertaking for an organization. It can seem overwhelming to anyone involved. Early wins can significantly help drive momentum. Then it is necessary to sustain those gains to continue the momentum. As discussed in chapter fourteen, a clear training plan should be developed. Starting with difficult parts will only lead to frustration. What differentiates a "simple" part from a "difficult complex" part? Simple parts would include additively manufacturing tooling and easy insertion opportunities. Conversely, if your company is new to AM and your company initially invests in an Arcam EBM machine, with no prior experience to produce all of your difficult complex parts out of AM, you will be in for a rude awakening.

Training should begin with simpler parts and simpler technology and move to more difficult parts once people within the company become more familiar with the technology. At first, senior leadership and the guiding coalition should carefully select parts for AM that are simple and will help train employees on the basics of AM. This will create an environment that promotes learning with little stress. Many companies will start with employees using AM to make fun designs such as their university insignia, favorite sports team, or other non-work-related objects. It is all about getting employees interested in the AM technology, so give a little room for some fun with the basic training. While it may not be making a part, it is beneficial to the change vision because it broke down a barrier. Then you can move into a work-related short-term win with tooling and fixtures, then difficult to source parts. This strategy is exemplified in Figure 16.1.

Short-term wins are often categorized as improvements that can be implemented in 6 months to a year. It is important that senior leadership

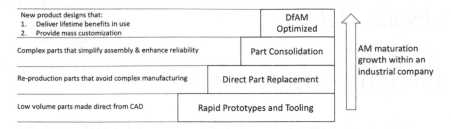

Figure 16.1 Renishaw Saunders' Staircase.

and the guiding coalition outline a multiyear plan with several short-term wins along the way. These short-term wins must be measurable and clear. People need to know if it was successful. Short-term wins must also be visible to all levels of the organization. Communicate the progress and the win so that everyone knows. Finally, the short-term win must be related to the change effort. If people cannot see the linkage, it will not benefit the AM adoption.

Every time a short-term win is accomplished, it needs to be celebrated. Details such as cost savings, improved customer satisfaction, and reduced lead time should be highlighted to show the impact of the AM technology. This drives the momentum of AM adoption. The communication also shows that this is a change that is staying, which will help sway any remaining resisters.

When looking for an initial short-term win, selective laser sintering of polymers and even material jetting may be used to learn some of the basics of laser powder bed fusion (L-PBF) of metals since they involve at least one of the key elements (laser or support structures). It is important to select a product or methodology that you have the internal skills and capabilities to be successful with.

Product selection matrix

Product selection can be one of the most difficult tasks during AM adoption since most organizations make numerous, if not hundreds, of different products. In the initial stages of AM adoption, it is important to select products that are not complex to aid in training and development of employees. If complex parts are initially selected, issues will likely arise in the AM process, which will result in resistance from employees. Therefore, simpler products should be selected first as part of an AM training program. To facilitate the product selection process, an AM steering committee can oversee this process by developing a well-defined and disciplined selection process.

The AM steering committee should be led by an executive-level member and consist AM experts, senior operations managers, material management, and logistics personnel. These individuals will bring experience to developing the product selection matrix in terms of feasibility and manageability.

It is not feasible for the steering committee to assess every single product initially. Therefore, the first step of the steering committee is to look at their product families for opportunities. Product families should be assessed based on their complexity and volume as shown in Figure 16.2. The initial focus (phase I) for product selection should be on simple, low-volume products as employees are trained in AM. This will enable training on the basic principles of AM without interrupting normal production operations and short-term wins. As more and more employees are trained, AM will start being adopted in the organization (phase II). In addition, the lead time and quality level will become more consistent and predictable. At this point, AM can be integrated into the daily production operations. In the third phase, the products considered for AM should become more complex. This focus will enable long-term wins. Training should also be aligned to continue to grow employee's AM skills. Depending on the types and varieties of materials used in AM and size of the organization, this phase may take a considerable amount of time. In addition, the financial impact of each part should be considered as the steering committee moves through the four phases.

The final products can then be assessed against each other using a product selection matrix as shown in Figure 16.3. The matrix enables the identification of product opportunities and expected benefits. In addition, existing AM products can be included as a reference to determine prioritization.

When selecting a product for AM, it is important to have a selection team to be cross-functional and have an understanding of AM. One of the

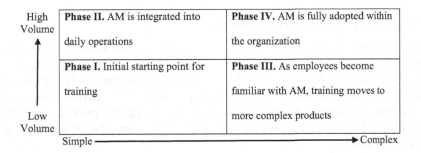

Figure 16.2 Product family assessment.

Product Selection Matrix					
Part Number		A	B	C	D
Product Description					
Selection Characteristic	Priority (1 = low, 10 = high)				
1.Will the product be production ready in six months?					
2.Will AM improve quality?					
3.Will the product lead-time be reduced?					
4.Will the product cost be reduced?					
5.Will the product strength be improved?					
6.Is the product complex?					
7.Will the product reliability be improved?					
8.Is AM the right approach for the product?					
9.Is success likely?					
Totals					

Figure 16.3 Product selection matrix.

downsides that can happen in these types of group scoring activities is when there is very uneven representation, say 6+ people from Structures and Manufacturing and only 1 or 2 from Design. This can over represent the opinions of a subgroup (which may tend to simply echo their leader's opinion) that may lack the expertise to appropriately assign and determine weighting factors in a scoring matrix, especially from a customer's perspective. One way to address this issue is to have weighting factors applied exclusively or at least primarily by those responsibilities most closely related to the judgment criteria. Another way to address this issue is to perform this process separately from the raw scoring, perhaps by a steering committee or more senior leaders responsible for these functions.

Pugh concept selection matrix

The purpose of a Pugh concept matrix is to evaluate several alternatives in order to best select the alternative against a certain set of specific criteria. It is a structured concept selection process used by multidisciplinary teams to converge on a decision using criteria. The methodology uses a matrix consisting of criteria and its relationship to specific, candidate products. Then, evaluations are made by comparing the products based on the criteria.

The uses of a Pugh concept selection matrix are to evaluate design concepts for a new product or process against a prioritized list of CTQs. Large decisions can be made within a Pugh matrix by evaluating alternatives

Criteria	Concept A	Concept B	Concept C	Concept n	DATUM
Criteria 1					
Criteria 2					
Criteria 3					
Criteria 4					
Criteria 5					
Criteria 6					
Criteria 7					
Criteria 8					
Criteria 9					
Criteria n					
Sum of (+)					
Sum of (-)					
Sum of (S)					

Figure 16.4 Pugh concept selection matrix.

in order to make important decisions such as make versus buy, supply chain operations, and purchasing of expenditures for capital investments. The Pugh matrix must be done as a team and should not be conducted alone since the results could be negatively skewed. Figure 16.4 provides the Pugh concept selection matrix.

The Pugh concept selection matrix is a structured approach that enables a team to converge on a superior concept. The matrix includes the criteria from the voice of the customer to evaluate candidate design concepts. The methodology is used to evaluate new concepts to an existing concept or standard, which is referred to as the "datum." The three main classification metrics include:

- Same as datum,
- Better than datum, and
- Worse than datum.

The process typically involves several iterations to combine the best features of the highly ranked concepts until one superior concept emerges. This superior concept then becomes the new datum or benchmark.

They key steps for the Pugh concept selection matrix include:

1. Determine the concept selection criteria using the voice of the customer
2. Determine the best-in-class concept that should be the datum from industry rankings or customer surveys
3. Develop 6–8 candidate concepts for evaluation

4. Evaluate each candidate concept against the datum using the selection criteria: (+), (−), and (S) are used to indicate better, worse, and same as the datum
5. Analyze the results of the evaluation by summing the number of (+), (−), and (S) ratings
6. Identify the weaknesses in the low-scoring concepts that can be turned into strengths
7. Identify the weaknesses in the high-scoring concepts that can be turned into strengths
8. Integrating the strengths of the concepts to create hybrid concepts
9. Add any new concepts or hybrid concepts
10. Re-evaluate using steps 4–8
11. Based on the re-evaluation, develop a single hybrid concept using the strengths of each concept
12. If necessary, repeat the steps for a third round of evaluation

Conclusions

The key decision is determining whether or not AM is appropriate for the product. If AM is suitable, then the steering committee must align the complexity of the product with the level of training, starting with simple and working to more complex products. Training employees on simple products will generate short-term wins and drive momentum toward an environment that adopts AM.

Concepts can be generated from several sources such as personal inspiration, vision, benchmarking, and reverse engineering. Once a team develops multiple potential design concepts, these can be evaluated against customer criteria using the Pugh concept selection matrix to determine the best candidate concept for AM. It's important to remember when selecting a project that the focus must be on short-term wins to gain and sustain the momentum.

chapter seventeen

Consolidate gains and promote additive manufacturing adoption

Once you've empowered broad-based action through basic system changes and created additive manufacturing (AM) awareness with training; AM projects should start to become visible and credible within the company. At this point, it is time to consolidate gains and promote more AM adoption. The concept is to leverage AM's increased credibility to drive change into larger scale systems, organizational structures, and bureaucratic policies that do not fit the vision. In addition, promote, hire, and groom employees who can implement the AM change vision. At this point, efforts should be made to examine the newly adopted processes with new AM projects that yield even greater opportunities for value. Next, let's break down these general guidance heuristics into actionable steps relatable to the AM industry.

A terrific way to begin this phase of change management driven by AM is by using the Lean Systems tool of value stream mapping (VSM) to understand current process structure barriers that would slow down AM adoption. VSM is a visual flowcharting tool that provides leaders with an approach to understand how their current system operates. This system depicted with VSM could be anything from design engineering, manufacturing, quality and inspection, purchasing, etc. Let's start with an example of a typical design engineering process cycle to see how detailing a VSM effort can assist in consolidating the gains for AM.

To start a VSM, one must be at least aware of Lean Systems identification of seven wastes of overproduction, waiting, transportation, overprocessing, inventory, and movement and defects. Originally created as wastes represented in manufacturing environments, these types of wastes can easily be applied to office environments as well as shown in Figure 17.1.

In our next example, we will use a case scenario of a typical design engineering department that has been tasked to implement AM into their product designs. The seven wastes of lean can be applied to this scenario as shown in Figure 17.2.

Now that we have defined a few examples of wastes involved in the AM design engineering department, we can start evaluating the current state process flow of how AM designs are driven through the traditional

Waste	Applied to Manufacturing	Applied to Office Environments
Overproduction	Producing things before demanded	Producing more than the customer needs or wants
Waiting	Inability to move to the next processing step	Idle time when people, material, information, or equipment is not ready
Transportation	Unnecessary movement of materials between processes	Transporting material or documents farther than necessary, or temporarily locating, filing, stocking, stacking, or moving materials, people, information, paper – wastes time and effort
Over-processing	Inappropriate processing of parts due to poor tool and/or product design	Effort that does not add value from the customer's perspective
Inventory	Storing more parts than the absolute minimum	More material information than the customer needs
Movement	Unnecessary movement of people during the course of their work	Bodily or mental motion that does not add value
Defects	Production of defective parts	Rework due to defects in the design or multiple design iterations.

Figure 17.1 Seven forms of waste.

Waste	Applied to AM Design Engineering
Overproduction	Creating 2D blueprint CAD drawings when AM parts are built from 3D models
Waiting	Wasted time waiting on CAD designers to become available for CAD changes between AM builds
Transportation	Designers constantly walking to and from the AM lab to assess their designs
Over-processing	Overanalyzing structural loading information, infinite optimization efforts
Inventory	Unnecessary building of 3D models that create desktop artifacts
Movement	Meetings to pontificate theoretical product usage scenarios that make up 0.0001% of use cases
Defects	Errors created in the CAD geometry, .stl file, poor support structure design

Figure 17.2 Seven forms of waste applied to AM.

design process established by the organization. Note that this process was originally designed to accommodate a more traditional manufacturing processes with many functional silos and not optimized for AM, but rather AM adapted to fit the established process. Value added (VA) indicates value added time in terms of days and NVA is non-value-added time in terms of days. The VSM basic architecture is noted in Figure 17.3.

Figure 17.3 is called a Current State Map, or CSM. A current state map is snapshot of how a process is currently working. By observing the figure, one can easily understand the process flow of information generated by various software systems all working to create a VA AM design. Some other observations from the VSM indicate a high amount of non-VA time waste in days associated with the sum of the processing steps. In total, there are between 13 and 26 days of wasted time in the process depending on if the parts produced met structural or dimensional requirements after initially produced.

The next step is where we evaluate the seven wastes determined above and start determining the root cause of all the non-VA waste in

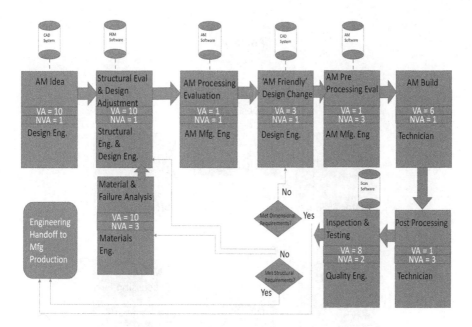

Figure 17.3 Basic VSM of the AM engineering process in a traditional organizational structure.

the process chain. Upon investigation, certain themes of waste are discovered. (1) Too many digital handoffs between software create a lot of "defect" waste. (2) Each of the engineers sits in separate buildings and with each process being very isolated, "transportation" waste starts to add up quickly. (3) The functionally siloed organization inherently contributes to "overprocessing," "waiting," and "movement" waste. These waste themes are amplified for AM processes due to the characteristic rapidly iterative nature of early design and manufacturing development.

Based on the issues surrounding 1–3, a brainstorming session is held to resolve the current structure to remove barriers inhibiting AM performance within the organization. The brainstorming session included a cross-functional group of individuals from many aspects of company and executive sponsorship to bring in many different perspectives. The outcome of that brainstorming session yielded a future state map. A future state map, or FSM, is your vision for the next 2–3 months; you define what you would like to improve and plan to achieve it within the timescale. Below, the team generated two major changes that could take place to change the process. (1) The first idea would be to look into enterprise solution software that accomplishes many of the software solutions under one system, which also would reduce existing costs. Siemens and Dassault are currently developing such AM systems. (2) The second idea was to

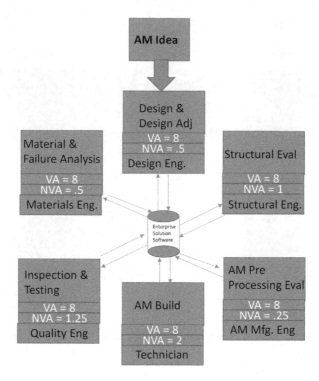

Figure 17.4 Proposed future state map to reduce process waste.

co-locate a cross-functional team to improve communications to reduce waiting wastes, reduce waste of motion and reduce waste of rework. The FSM now looks like Figure 17.4, with NVA being reduced to 5.5 days.

To recap, the above example is to provide the reader with a concept of how to use the VSM tool to clearly identify barriers in your processes and how to parse out solutions that drive change into larger scale systems, organizational structures, and bureaucratic policies that don't fit the vision. For brevity, the above example is actually a watered down version on how to properly accomplish VSM. The authors recommend having a Lean consultant work with your organization to map out a more detailed VSM that encompasses several departments and provides a broader picture of the organizational systems involved in adoption AM within the company.

Assuming that you lead several VSM projects discovering structural or procedural barriers that prevent AM adoption, efforts should now pivot to examine the newly adopted processes with new AM projects that yield even greater opportunities for value. But how does one go about evaluating which proposed AM projects yield more value than others? Like the saying, "there's an app for that." Similarly, there is a "Lean Six Sigma (LSS) tool for that." The PICK chart is an exceptional way to evaluate which project may or may not yield a value for AM and is "used to categorize process improvement ideas during an LSS event. The purpose is to help identify the most useful ideas. The 2×2 grid is normally drawn on a white board or large flip chart. Ideas that were written on sticky notes by team members are then placed on the grid based on the payoff and difficulty level. The acronym comes from the labels for each of the quadrants of the grid: Possible (easy, low payoff), Implement (easy, high payoff), Challenge (hard, high payoff), and Kill (hard, low payoff) as shown in Figure 17.5.[1]

The most appropriate way to generate a PICK chart and evaluate AM ideas is in a brainstorming session with input from functional leaders as well as engineers, technicians and everyone who has ownership in the proposed AM solutions. Keep in mind that it doesn't matter "exactly" where each idea will fall on the chart, but rather relative orientation to each quadrant. The ideas fall in the implement category should be the first ideas to start with. Also, discourage anyone placing ideas on the lines themselves as this will only encourage a team to delay making decisions. By having a team of AM shareholders present and providing input into the PICK chart, they will have more ownership in the process and, as a result, be more successful in adopting the AM project ideas in the company.

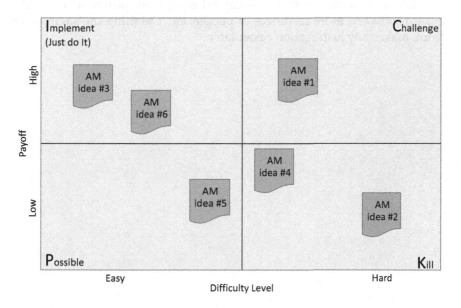

Figure 17.5 PICK chart example.

[1] www.vertex42.com/ExcelTemplates/PICK-chart.html.

Nothing provides more justification for AM than recorded value assessments of the AM successful projects. It is critically important that as one goes through the process and rigor of recording value saved from AM implementation. Often, the metrics associated with this practice would include, development lead time savings (days), cost avoidance ($'s), direct cost savings ($'s), cost of poor quality from conventional process ($'s), and weight saved through AM optimization (lbs). If implementing a new AM design, and you have integrated many parts into a single AM design solution, more lifecycle costs could also be considered. Though showing up mainly as cost avoidance classification, these may include, assembly and transportation savings ($'s), warehousing and storage ($'s), disposal and recycling cost savings ($'s), and improving quality to affect warranty costs ($'s). On the manufacturing side, there have also been research studies conducted that highlight aggregate energy savings from using AM metals versus conventional manufacturing, though admittedly, this would reveal itself in the direct fab cost comparison. In term of the format to showcase these costs, there is no exotic Six Sigma/Lean template to showcase these savings, just a simple excel spreadsheet filled out cooperatively with your finance team should suffice.

Once, you have detailed the value opportunity of introducing AM into the company through the above financial approach, you will find it much easier to justify that new capital equipment purchase, new AM design software, more resources of people, etc. Detailing the savings up front, makes any justification easier later.

chapter eighteen

Anchor new approaches in the culture

An organizational culture may be described as a set of general behaviors, ways of thinking and shared values. These cultures are not static, but continuously changing over time in response to the broader environment and transformations in ways of working. In this book, we have illustrated the many ways that additive manufacturing (AM) is a transformer of culture that can be powerfully positive and make your organization a more exciting and energizing place to work. However, there are also potentially negative cultural elements that come with such a disruptive change and an experienced leader will recognize how this can inhibit growth, damage morale, and destroy trust between your team members. The process restructuring required to realize the benefits of AM must be anchored in the culture to avoid regression to the initially more comfortable "old ways."

It has been this author's experience that adopting and anchoring AM change within the company culture is harder than it may seem. I have witnessed and heard about many companies taking the time to communicate the benefits of AM within leadership, generated new AM initiatives, purchased AM machines, identified and sourced AM talent, trimmed dead weight employees and obstructionists that resist AM and generally work to achieve the AM implementation plan set forth many years prior. These companies have picked the low hanging fruit AM projects that augment the AM initiatives that align to the strategic plan and implemented them successfully. They have tracked and metricized all the right AM attributes of lead time shortening, tooling cost savings, design cycle reductions, etc. The leadership within the companies begin to congratulate each other, some even retire knowing that their work industrializing AM within the company is complete.

However, something seems wrong. The AM implementations are remaining isolated. Management seemed baffled. Why would this be the case? Executive management learned how to apply the technology and allocated the funding to do so. Why isn't AM adoption sticking within the culture? The answer is a relatively simple one and is often overlooked within companies trying to transform their products using AM. The new technology isn't spreading throughout the organization simply because the folks at the worker level (the people affected by AM implementation the most) had little say into what AM projects to do or how to apply the technology. They were "told" to implement AM, they were not "empowered" to deploy AM.

The last step of Kotter's change management model is about anchoring the change within the organization. For AM, this is a culmination of all the hard work the company has put into creating cultural change within the organization. At this stage in the change management process, the company must take stock in all of the AM change that has taken place thus far and compare it to corporate success. In addition, it is during this phase of change, that the company must ensure an AM legacy through leadership development, management succession planning and perhaps most importantly, worker-level training workshops.

So, what tool can be used to drive strategic goals associated with AM throughout all aspects of a company? Lean system best practices teach us that Hoshin Kanri, aka, policy deployment, is a methodology that ensures efficient communication, strategic direction and goal alignment is aligned, thereby created a shared vision throughout "all" levels of the company. In other words, this ties the entire company together to ensure that new approaches to AM will be anchored within the company.

The history of Hoshin Kanri is interesting and was largely held as business best practice for most of the 1980s and 1990s in the United States. The Japanese are credited with developing the management philosophy of Hoshin Kanri. In short, the Japanese mixed Juran's and Demings' instruction with Peter Drucker's writing on Management by Objectives and created the idea of strategic quality planning. Each individual company created their own planning processes. "In 1965, Bridgestone Tire published a report analyzing the planning techniques used by Deming Prize winning companies. The techniques described were given the name Hoshin Kanri. By 1975, Hoshin was widely accepted in Japan.[1]"

The basic premise of Hoshin Kanri is quite simple. The first step is to develop a strategic plan. This has been covered in detail in Chapter 13. The second step involves developing tactics. This has been referenced in chapters fourteen and fifteen. The third step is taking action within an organization which has been covered in chapters sixteen and seventeen. The fourth step is review and adjust, which brings us to our current chapter, anchoring new approaches for the AM change. One way to do this is through Standardized Work.

Taiichi Ohno, father of the Toyota Production System, has been commonly attributed as saying "When there is no standard, there is no Kaizen," while others have attributed it to Masaaki Imai as "Where there is no standard, there can be no improvement [1]." Regardless of the originator, the

---

[1] http://mcts.com/hoshin-history.htm.

principle is critically important to sustaining change. Standardized Work is a discipline in which teams document the current practice, which provides a clear basis for ongoing or continuous improvements. There are many benefits of Standardized Work discussed in the literature, but organizations attempting to solely drive AM adoption top-down will struggle with compliance long-term; as we discussed in chapter fourteen, broad-based action is required.

Reducing defects and waste is a primary benefit of Standardized Work to anchoring change. Many teams new to AM become infatuated with "going fast" when exposed to the new capabilities. Process-oriented leadership is critical in the early stages of your AM implementation. Without making the time to establish standards, your team will invariably waste time duplicating work when a new project comes around. Even worse, work necessary to deliver a high-quality AM product is likely to be missed in the up-front cost or time to complete estimates. This will result in disappointed and frustrated customers wondering why a supposedly rapid process seems to take so long, as they experience repeated delays, suboptimized results, and higher project and product costs than promised.

However, it is very difficult to turn around a negative opinion once entrenched and will require more work than if the job was done well in the first place. Much like a terrible restaurant experience spreads within social networks, a poorly executed AM project will end up being a long-term roadblock no matter how much "top-down" pressure to adopt AM is exerted. "Going (too) fast" can hinder credibility and set back your efforts to anchor necessary changes in the engineering culture.

With AM, this resistance typically manifests through project managers highlighting the risks to schedule, production costs, and "unknown-unknowns" and eventually recommending parallel pathing of conventional vs. additive design. Performing parallel path efforts for multiple projects will be costly to your organization and is especially disastrous for an additive approach. Given the typical resource constraints, spending this extra effort further reinforces the negative opinion of teams competing for the same pool. It may make it impossible – or at least too costly – to perform DfAM at a system level as changes must accommodate both conventional and AM technology, limiting the performance improvement possible. Lastly, it also ties your agile AM projects to a conventional timeline, potentially negating one of the primary benefits. In the end, the program manager's prophecy becomes self-fulfilling when an additive design is delivered on a slow timeline, at not much better performance, with high costs.

A secondary benefit of Standardized Work to anchoring AM in your culture is that it simplifies hiring, onboarding and training. If you have a clear understanding of the required work and have documented these required steps, then development of a training and onboarding plan with AM relevant skills and understanding at its heart becomes straightforward. A comprehensive understanding of the involved processes through value steam mapping and other tools will also identify skill gaps, and desirable mindsets, for targeted hiring and cross-training.

Rather than being frustrated by vague assertions to "think addively," new employees will understand expectations and can begin the hard work of actually doing additive. This is a key element of sustaining the AM-driven cultural changes in the broader organization. Establishing the standard with those performing the work and reinforcing it with strong accountability and technical leadership from experts makes it possible to both get effective results time after time and to continue to get better. Instead of being an isolated team of experts, AM will be able "scale" within your entire organization.

Much like AM itself, a third way that Standardized Work can anchor change in your culture is that it, in fact, encourages creativity and flexible thinking. Suggestions to continuously improve current processes and optimize designs are actively solicited, a key element and self-reinforcing feature of the desired AM-enabled engineering culture. Contrast this with a typical pedigree-focused engineering design approach, which strongly prioritizes what has been proven to work before, until forced to try something new due to evolving requirements. While it has a place, we have personally battled design review checklist questions asking "Have all features of the design been selected from previous designs?" and "Are only common parts from [Approved List] selected?" Most teams will eventually progress past this point, but we have seen it exploited effectively by resisters of change. Showing the Standardized Work and process rigor behind a specific AM proposal is an essential tactic for breaking down this type of barrier and converting resisters into promoters.

The strongest anchor for a shift in an organization's culture is delivering results – ways of working that directly enable success are contagious. Following the guidelines and rigorously applying the tools highlighted in the preceding chapters will improve the probability of consistent, high-impact wins from your AM efforts. This will help you build momentum in your organization and guide it toward lasting transformation.

If you remember anything from reading this book, please remember the following. Introducing a new technology, such as AM, is difficult. It is incredibly challenging and rewarding at the same time. There will be cynics. There will be obstructionists. There will be people who will sabotage any AM efforts to curtail AM adoption. This is largely a symptom of the technology itself being disruptive. Disruption is perceived to be dangerous, and it can illicit strong feelings of livelihood risk and fear.

However, at the same time, AM is exciting and it offers tremendous potential to engineer products beyond how we have done it in the past.

"Please," do not give into the barriers of AM change. Be an advocate. Be a leader. Be courageous and passionate about adopting new technologies. Most of all, you need to remember that change must occur to keep your company profitable.

As you enter your AM journey within your company, we will leave you with a quote by Charles Darwin. "Ignorance more frequently begets confidence than does knowledge: it is those that know little, and not those who know much, who so positively assert that this or that problem will never be solved by science."

Reference

1. Imai, M. (1986). *Kaizen: The Key to Japan's Competitive Success* (1st ed.). New York: McGraw-Hill Education.

Index

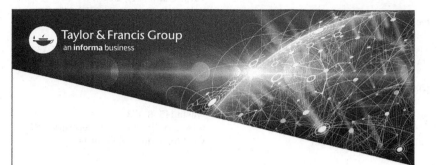

Printed in the United States
by Baker & Taylor Publisher Services